Adapting High Hydrostatic Pressure (HPP) for Food Processing Operations

Adapting High Hydrostatic Pressure (HPP) for Food Processing Operations

Tatiana Koutchma, PhD
Agriculture and Agri-Food Canada
Guelph Food Research Center
Guelph, ON, Canada

AMSTERDAM • BOSTON • HEIDELBERG • LONDON
NEW YORK • OXFORD • PARIS • SAN DIEGO
SAN FRANCISCO • SINGAPORE • SYDNEY • TOKYO
Academic Press is an imprint of Elsevier

Academic Press is an imprint of Elsevier
32 Jamestown Road, London NW1 7BY, UK
525 B Street, Suite 1800, San Diego, CA 92101-4495, USA
225 Wyman Street, Waltham, MA 02451, USA
The Boulevard, Langford Lane, Kidlington, Oxford OX5 1GB, UK

First published 2014

Notices
Knowledge and best practice in this field are constantly changing. As new research and
experience broaden our understanding, changes in research methods, professional practices, or
medical treatment may become necessary.

Practitioners and researchers must always rely on their own experience and knowledge in
evaluating and using any information, methods, compounds, or experiments described herein. In
using such information or methods they should be mindful of their own safety and the safety of
others, including parties for whom they have a professional responsibility.

To the fullest extent of the law, neither the Publisher nor the authors, contributors, or editors,
assume any liability for any injury and/or damage to persons or property as a matter of products
liability, negligence or otherwise, or from any use or operation of any methods, products,
instructions, or ideas contained in the material herein.

Library of Congress Cataloging-in-Publication Data
A catalog record for this book is available from the Library of Congress

British Library Cataloguing-in-Publication Data
A catalogue record for this book is available from the British Library

ISBN: 978-0-12-420091-3

For information on all Elsevier publications
visit our website at **http://store.elsevier.com/**

CONTENTS

CHAPTER *1*

Introduction

The benefits of hydrostatic pressure processing (HPP) technology are briefly introduced for the range of food applications. Geography of HPP products market and products categories are discussed along with the perspectives of the HPP equipment manufacturing. The need in practical recommendation and steps is justified to accelerate the HPP technology transfer.

1.1 HPP BENEFITS AND GLOBAL MARKET

High hydrostatic pressure processing (HPP) is a novel technique that emerged in food production within the last decade and more recently for pharmaceutical and medical applications. Key advantages of HPP such as instant and homogenous transmittance of pressure in the treated product and the possibility of processing at ambient temperatures represent an attractive powerful tool to implement mild postpackaging processing for a broad range of foods. Pathogens inactivation and shelf-life extension in ready-to-eat (RTE) meats are, apparently, the HPP applications that have made the technology known and accepted in the food industry. Theoretically pressure–temperature operational conditions for HPP treatment of foods may range from 300 to 1000 MPa at $-20°C$ to $121°C$, depending on the process and the nature of the food resulting in the processes such as sterilization, pasteurization, or shelf-life extension with added value.

HPP as a clean label technology has found a growing acceptance in the food industry for producing high-quality foods. The recent market research study reported that by 2013 the global HPP products market has achieved $3 billion. Fruit and vegetable products are the largest product market for HPP, followed by meat products and then seafood and fish products. North America dominates the global HPP products and equipment market, with the United States as the largest market. United States as a leading adaptor of the technology is followed by parts of Europe (Spain, Germany, France, The Netherlands, Great Britain, and Italy), Asia (Japan, China, Taiwan), Australia, and New

Adapting High Hydrostatic Pressure for Food Processing Operations. DOI: http://dx.doi.org/10.1016/B978-0-12-420091-3.00001-7

Zealand. In 2013, the estimate of HPP equipment market was $350 million and is projected to grow at 11% Compound Annual Growth Rate (CAGR) for the next 5 years.

Vast amount of knowledge and information was generated in the area of HPP research and knowledge transfer since Hite discovered and first studied pressure effects more than 100 years ago. In 2012, the Food Safety Working Group (FSWG, http://www.cigr.org/governance_work. html#FoodSafety) of CIGR in collaboration with Guelph Food Research Centre of Agriculture and Agri-Food Canada (AAFC) conducted a survey to collect the answers and opinions of food professionals in terms of the role that novel and emerging food processing technologies including HPP can play to address global food safety issues and challenges and the level of applications. The answers were received from 87 respondents in North and South America, Europe, Asia, New Zealand and Australia, and Africa. Among 11 novel processing technologies, the respondents of the survey selected the following commercial application that emerged in food production in their countries: microwave heating was highlighted by 88%, high pressure by 80%, and UV light by 84% of total respondents. When asked about the technologies which have potential to be commercialized in 5–10 years, the top three technologies selected were HPP, microwave heating, and UV light. When answering about main drivers of innovations, equipment manufacturers (73%), academia, and government research organizations (82%) as well as large food corporations (81%) were selected as the three key players who contribute in innovation in the food industry. Equipment manufacturers and communication and knowledge transfer media also scored above 50%.

With growing adaptation of HPP technology, there is a need along with understanding of fundamental principles, to give practical recommendations on how HPP products can be developed and HPP processes can be applied to specific operations considering available scientific information. This includes phenomena of compression heating, pressure effects on biological organisms and food matrixes, microbial efficacy in different food matrixes, product characteristics, impact on quality parameters, and suitability of packaging materials. At the same time, the effect of critical processing parameters that define pressure treatment cycle in combination with temperature have to be better understood from efficacy and productivity point of view depending on the application.

This publication is intended to discuss common technical principles and commercial benefits of HPP as a rapidly emerging novel processing technique. Fundamental principles and HPP mode of operation along with the critical considerations for process development and selection of commercial equipment will be discussed. Commercial benefits of HPP technology are presented for specific product categories that include applications for raw and RTE meats, dairy and seafood products, fruit and vegetables beverages, and other emerging processes. Additionally, energy requirements and regulatory status of HPP technology around the world including United States, Canada, EU, and Asia will be updated. Thus, this publication will take a look on how the results of the research in HPP technology can be successfully converted to commercial preservation and value-added processes and novel food products with extended shelf-life and result in new profit opportunities.

CHAPTER 2

Fundamentals of HPP Technology

This section discusses fundamental HPP principles, range of elevated pressures and temperatures used for food applications and existing methods of pressurization. The pressure effects on food chemistry and other essential food properties under the pressure such as phenomena of adiabatic compression, compressibility, thermal expansion, conductivity, and specific heat are reviewed using specific examples for different food categories.

2.1 HPP PRINCIPLE

High HPP or high pressure processing utilizes elevated pressures typically in the range of 100–1000 MPa with or without addition of external heat. Most often used pressure units and their conversion are listed in Table 2.1.

In typical batch HPP operation, prepacked food is loaded inside a pressure vessel and is then pressurized by utilizing water as a pressure-transmitting medium. With direct compression used in lab scale systems, the volume of the vessel is reduced by the action of a hydraulic pressure applied to a piston. In indirect compression systems, an intensifier, or high-pressure pump, is used to pump a pressure-transmitting fluid directly into the vessel to achieve a target pressure. The indirect method of pressurization is currently used for the commercial application of HPP in food processing. The volume reduction of water and water-based foods at 600 MPa and room temperature is approximately 15% (Bridgman, 1912).

2.2 FOODS UNDER PRESSURE

The governing principles of HPP processing are based on the assumption that foods under high pressure in a vessel follow the isostatic rule regardless of their size or shape. According to this rule, pressure is instantaneously and uniformly transmitted throughout a sample, whether the sample is in direct contact with the pressure medium or hermetically sealed in a flexible package. Therefore, in contrast to

Adapting High Hydrostatic Pressure for Food Processing Operations. DOI: http://dx.doi.org/10.1016/B978-0-12-420091-3.00002-9

Table 2.1 Pressure Units				
	Atmosphere (atm)	Bars (bar)	Megapascal (MPa)	Pounds per Square inch (psi)
1 atm	1.000	0.987	9.901	0.068
1 bar	1.013	1.000	10.000	0.060
1 MPa $(1 Pa = 1 N/m^2)$	0.101	0.100	1.000	0.00689
1 psi	14.696	14.504	145.038	1.000

thermal processing, the time necessary for HPP processing should be independent of the sample size.

The effect of HPP processing on food chemistry and microbiology is governed by the Le Chatelier's principle. High pressure stimulates some phenomena (e.g., phase transitions, chemical reactions, and changes in molecular configuration) that are accompanied by a decrease in volume but opposes reactions that involve an increase in volume. The effects of pressure on protein stabilization are also governed by this principle, i.e., the negative changes in volume that occur with an increase in pressure cause an equilibrium shift toward bond formation. Alongside this, the breaking of ionic bonds is also enhanced by HPP, as this leads to a volume decrease due to the electrostriction of water. Moreover, hydrogen bonds are stabilized by high pressure, as their formation involves a volume decrease. Pressure does not generally affect covalent bonds. Consequently, HPP can disrupt large molecules and microbial cell structures, such as enzymes, proteins, lipids, and cell membranes, and leave small molecules such as vitamins and flavor components unaffected (Linton and Patterson, 2000).

2.2.1 Adiabatic Heat of Compression

During pressurization, the packaged product undergoes isostatic compression by the pressure-transmitting fluid, causing a reduction in the package volume (compressibility) of up to 19%, depending on the final pressure and temperature reached. Fluids such as air, water, and food materials undergo adiabatic heating during compression and cooling during decompression.

Adiabatic heat of compression is the instantaneous volumetric temperature change in materials during compression or decompression is often called the heat of compression. Equation (2.1) expresses this adiabatic heating:

$$\frac{dT}{dP} = \frac{T\alpha_p}{\rho C_p}$$ (2.1)

As indicated by Equation (2.1), the temperature increase dT depends on the volumetric expansion coefficient α_p (measured in units of K^{-1}), the density ρ (kg/m^3), and the isobaric heat capacity C_p (J/kg/K) of the materials well as on its initial temperature T (K).

The quasi-adiabatic temperature increase δ_S upon compression of food samples can be evaluated as the ratio of the temperature increase to the pressure increase using Equation (2.2):

$$\delta_S = 100 \times \frac{\Delta T}{\Delta P} = 100 \times \frac{T_f - T_i}{P - P_{atm}} \cong 100 \times \frac{T_f - T_i}{P} \left[\frac{^\circ C}{100\ MPa}\right]$$ (2.2)

where T_f is the temperature of the sample at the applied pressure, T_i is the initial temperature, and P is the applied pressure. Water, carbohydrates, fats, and proteins are the main components of the complex food matrix that respond uniquely under compression. It has been reported that water has the lowest rate of temperature increase under compression, about 3°C per 100 MPa at 25°C, whereas fats have the highest value, up to 6.7–8°C per 100 MPa (Barbosa-Cánovas and Juliano, 2008; Rasanayagam et al., 2003; Ting et al., 2002). Nevertheless, only limited information is available on the temperature rise with compression of real foods with complex fat compositions.

As follows from Equation (2.1), under compression heating conditions, the sample temperature at process pressure is dictated by the sample's thermodynamic properties. In reality, heat transfer between the sample and the pressure-transmitting fluid or the pressure vessel during the come-up time and hold time will likely influence the measured sample temperature.

As follows from Table 2.2, significant differences in the compression heating behavior were observed in foods with high oil/fat content such as vegetable oil, cheese, and mayonnaise, and foods with high water content such as milk. Foods with a high water content such as milk experienced a temperature increase with pressure change similar to that of water (the polar component), in the range of 3°C per

Table 2.2 Compression Heating (°C/100 MPa) Factors of Selected Food Substances Determined at Initial Product Temperature of 25°C (Patazca et al., 2007.)

Food Substances	Temperature Increase (°C/100 MPa)
Orange juice, tomato salsa, skim milk, salmon fish	2.6–3.0
Carbohydrates	2.6–3.6
Proteins	2.7–3.3
Mayonnaise	5.0–7.2
Extracted beef fat	6.2–9.1
Olive oil	6.3–8.7

100 MPa. Another pattern of behavior under HPP was found for the group of foods with significant proportions of nonpolar components such as cheese, mayonnaise, and vegetable oil. The magnitude of the temperature increase decreased with increasing pressure for oil and mayonnaise and did not change significantly for cheese. Similarly, the temperature increase for foods with a high content of proteins such as meat (beef and chicken breast) did not deviate significantly from that of water-like products. Experimental δ_S values for oil/fat and water obtained in this study can be used as reference values.

After the pioneering work of Bridgman (1931), the properties of water under pressure were well documented. Data are available from the International Association for the Properties of Water and Steam (IAPWS). A software implementation of IAPWS work can be obtained from the US National Institute of Standards and Technology (NIST). Most high moisture foods are assumed to have thermodynamic properties similar to those of water at the pressures and temperatures used in HPP. This assumption is less valid for lower moisture, high solids, and nonuniform systems. Very limited information is available on properties of food materials under pressure because of the practical challenges associated with the *in situ* measurement of these properties at elevated pressures (Ramaswamy et al., 2005).

2.2.2 Compressibility

Compressibility is an intrinsic property of the material and is defined by the balance between attractive and repulsive potentials. Compression of a liquid decreases the average intermolecular distance and tends to reduce rotational and translational motion. Food material (e.g., orange juice) is considered to contain molecules that occupy space in excess of that needed for close packing. This excess is called "free volume" and it is this volume that is reduced in initial compression (Rasanayagam et al., 2003). At elevated pressures, when the free volume has largely disappeared, reduction in van der Waals dimensions may occur and the compressibility is greatly diminished. Isothermal compressibility (β) is defined as the relative change in volume (V) with pressure (P).

$$\beta = -\frac{1}{V}\left(\frac{dV}{dP}\right)P \qquad (2.3)$$

Compressibility and density were reported by Min et al. (2010) for 16 foods based on volumetric measurements at 25°C and pressures up to 700 MPa. This included sucrose solutions (2.5–50%), soy protein solutions (2.5–10%), soybean oil, chicken fat, clarified butter, chicken breast, ham, cheddar cheese, carrot, guacamole, apple juice, and honey. Compressibility decreased with pressure at a rate that decreased with increasing pressure. For equal mass concentrations, protein solutions were less compressible ($p < 0.04$) than sucrose solutions at pressures greater than 200 MPa. Relative to water, fats showed high compressibility to 100 MPa, similar compressibility from 100 to 300 MPa, and less compressibility from 300 to 700 MPa. Chicken breast, ham, cheddar cheese, carrot, and guacamole showed relatively large compressibility from 0.1 to 100 MPa. Honey showed the smallest volume decrease over 700 MPa. Variability in compressibility of different materials is likely due to differences in concentration, chemical composition, and complex interactions between components within a food system.

2.2.3 Thermal Expansion

Thermal expansion of materials is caused by temperature change during compression. Thermal expansion coefficient (α) is another thermodynamic property which provide a measure of the amount by which the density change in response to a change in temperature at constant pressure (P):

$$\alpha = \frac{1}{V}\left(\frac{dV}{dT}\right)T \qquad (2.4)$$

Estimation of volume (Equation (2.3)) or thermal expansion (Equation (2.4)) under pressure was constrained by practical challenges associated with developing reliable sensors and other instrumentation that can withstand elevated pressure conditions.

2.2.4 Thermal Conductivity

Very limited studies reported thermal conductivity (k) of food materials under pressure. In general it was observed that material thermal conductivity increased with increase in pressure. It was influenced by the amount of moisture present in the food material. Water and water-like substances (apple juice) were found to have the highest k values (up to 0.82 W/m°C at 700 MPa and 25°C), while fatty foods such as canola oil and clarified butter had the lowest (0.29–0.4 W/m°C, respectively, at 700 MPa and 25°C) values. Honey and high fructose corn syrup had intermediate values. The pressure dependency of thermal conductivity values was higher at lower moisture contents than at higher moisture contents. This may be due to the smaller percentage of incompressible water and higher porosity of the sample at lower moisture contents. The estimated k values of all the food materials tested under pressure were higher than corresponding k values of materials under atmospheric pressure signifying faster thermal equilibrium under pressure (Ramaswamy et al., 2007).

2.2.5 Specific Heat

The data on specific heat of pure water as a function of elevated pressure and temperature are readily available through the NIST database (Harvey et al., 1996). These values were approximately 10% lower than those estimated at atmospheric pressures. In the absence of experimental data, researchers often ignore the effect of pressure on specific heat in heat transfer calculations.

2.2.6 Density

When food materials are processed under pressure, significant change in the volume of the product occur possibly due to reduction in the "free volume" between molecules or the compactness of voids occupied by air. The density of materials under pressure can be estimated by knowing the change in volume of materials under pressure and its mass. Density is then calculated as the ratio of mass to volume. The volume change of materials under pressure can be estimated by using linear velocity differential (LVD) transducer (Bridgman, 1931). Min et al. (2010) reported density increase of 16 food samples at a decreasing rate. Density under pressure for foods with relatively high solids content, fat content, and porosity deviates from the behavior of water, which is often used as an input in temperature distribution simulations.

CHAPTER 3

In-Container HPP Principle

The packaging aspect of HPP is discussed. The requirements for plastic packaging materials and their compatibility with HPP technology are detailed along with the packaging design and a brief review of reported studies.

Packaging is a vital factor for any preservation method to be successful since it protects the food products from recontamination during processing and adverse environmental conditions during storage. HPP is a batch process in which food materials are packaged before the application of high pressure, and the food product and the package are both exposed to the processing conditions, so the entire pack remains a "secure unit" until the consumer opens it. Prepackaging helps to prevent contamination from the pressure medium and improves HPP efficiency. However, any damage or alteration in the package or packaging materials can potentially cause the loss of its hermetical properties. Therefore, quality, safety, and shelf-life of the food product can be adversely affected.

In-container processing requires packages in the form of pouches, large bulk bags, or container−lid combinations; 90% of HPP foods are expected to be processed in flexible or partially rigid packages.

The packaging aspect of HPP is very essential for the successful application of the technology, and a number of studies have been conducted in recent years regarding the effect of high-pressure processing on packaging materials and food−packaging interactions (Bull et al., 2010). The severe processing conditions used in high-pressure applications may cause changes in the structure of packaging materials and therefore the barrier and mechanical properties of the package may be altered. When high-pressure and -temperature combinations are used in order to achieve sterilization effect, prepackaged food products need to be preheated to a targeted temperature before high pressure is applied. This reasonably harsh preheating process also has the potential to further affect the properties of the packaging material used (Koutchma et al., 2010). Metal cans, glass bottles, and paperboard are not suitable for the process.

Adapting High Hydrostatic Pressure for Food Processing Operations. DOI: http://dx.doi.org/10.1016/B978-0-12-420091-3.00003-0

Research related to the effects of HPP on food quality and package has been mostly conducted either at room or moderate temperatures. It has been shown that high pressure itself does not have considerably negative effect on properties of the packaging materials commonly used in food industry. However, if the adhesion between the layers of multilayer structure of the packaging film is affected during processing, gaps may appear within the structure resulting in loss of integrity.

Similar to food products, packaging materials also undergo compression heating. Knoerzer et al. (2010) have shown that polypropylene (PP) and polyethylene (PE) undergo compression heating greater than water under both HP and low and mild temperatures of 10°C and 50°C and HPP–high-temperature (90°C) conditions up to 750 MPa. In particular, the temperature increase with pressure was not linear and, therefore, the relative increase with respect to water depended on the pressure range selected as well as the initial temperature. Since in typical industrial process 5% of HP vessel is filled with packaging, compression heating behavior of packaging materials may affect heat transfer by forming insulation layer surrounding food and consequently temperature distribution within HP vessel.

3.1 REQUIREMENTS FOR PLASTIC PACKAGING MATERIALS

3.1.1 Flexibility

A packaging material needs to be flexible enough to withstand compression forces so that it can avoid irreversible deformations and maintain its physical integrity. Typically, high barrier flexible pouches made of polymers or copolymers with at least one flexible side can be used for processing solid or liquid food products.

3.1.2 Package Design

Package size and shape are also critical in terms of maximizing the number of packages, which can be fitted in the chamber. Therefore, proper package design can contribute to the economical processing.

3.1.3 Integrity

Packaging materials used for HPP need to endure harsh treatment conditions so that they can sustain the physical integrity both during the process and extended storage. Maintaining the integrity during both the process and storage is particularly important in terms of the safety and quality of the food products. Ideally, as well as maintaining its mechanical and other built-in features, packaging material needs to protect consumer acceptance and convenience (Lambert et al., 2000). Barrier, mechanical, and mass transfer properties (sorption and migration) of the package must be resistant to the changes occurring during the process. During the process, temperature increases and the volume of the package decreases due to the pressure applied. Once the pressure is released, the package needs to regain its original conditions.

3.1.4 Barrier Properties
Packaging material needs to have desired barrier properties to maintain the quality of food products especially during the shelf-life. Water and oxygen are two of the most important components in food which determine the quality and shelf-life. Barrier properties of the polymers are very vital considerations for shelf-life and type of polymer chosen plays a crucial role at this point. Developing packaging materials with improved water, oxygen, and light impermeability properties is essential to ensure desired shelf-life for high pressure and pressure-assisted thermally processed food products. Sealing is an important point for flexible pouches, and seal strength should be maintained during the processing in order to avoid any leakage (Koutchma et al., 2010).

3.1.5 Minimum Headspace
Vacuum packaging is important for a uniformity of HPP treatment because air in the package has higher compressibility. Excessive amount of gases can increase the come-up time of the processing. Vacuum packaging can also help avoiding oxygen-related reactions including lipid oxidation during both processing and storage. Vacuum packaging can improve loading factor, and more packaged product can be processed at one time in the limited volume of the pressure vessel.

3.1.6 Compatibility of Packaging Materials and HPP
Most of the flexible packaging materials used in the food industry show reasonably good compatibility with high-pressure processing when there is no severe heat contribution. Single and combinations of PET, PE, PP, and EVOH films are some of the commonly used packaging materials for HPP pasteurization (Juliano et al., 2010). Additionally, coextruded films with polymeric barrier layers, adhesive

laminated films on a polymer base, or inorganic layer such as aluminium foil or more recently vacuum deposited coating are used. Polyvinylidene chloride (PVDC), ethylene vinyl alcohol (EVOH), poly-vinyl alcohol (PVAL), and polyamide (PA) are common high barrier polymers.

CHAPTER 4

HPP Cycles

The section is focused on HPP process parameters for operations of pasteurization and sterilization. The differences in HPP cycles are discussed. The definitions of HPP parameters are given with the detailed effects of the critical product parameters and ranges of their variations. The impact, importance of proper documentation, and control in regards to HPP efficacy is discussed.

HPP utilizes elevated pressures typically in the range of 100−700 MPa with or without addition of external heat.

4.1 HIGH-PRESSURE PASTEURIZATION

Pasteurization treatment typically employs pressures in the range of 600 MPa (87,000 psi) at or near ambient temperatures (high-pressure−low-temperature process, HT−LT) for a specific holding time. Inactivation of pathogenic vegetative bacteria is the primary objective and product requires refrigeration storage.

Pressure-assisted thermal processing (PATP) or pressure-assisted thermal sterilization (PATS) sterilization treatment use elevated pressures (500−900 MPa) combined with a several minutes heat (90−121°C) exposure at pressure to sterilize low acid foods (high-pressure−high-temperature process, HP−HT). Inactivation of bacterial spores is the primary objective and product normally stored at ambient temperature.

4.2 PROCESS PARAMETERS

Figure 4.1 shows schematically the examples of HPP temperature and pressure cycles during HP−LT and HP−HT processing utilized for pasteurization and sterilization, respectively. The duration of the processing cycle shown in the figure is specific to the design of HP vessel used.

Adapting High Hydrostatic Pressure for Food Processing Operations. DOI: http://dx.doi.org/10.1016/B978-0-12-420091-3.00004-2

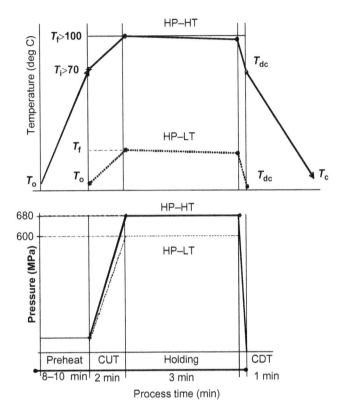

Figure 4.1 HPP temperature and pressure cycles.

When describing the HPP process, the proper documentation of HPP conditions is important and should include the following descriptors.

4.3 TIME

Cycle time: The total time for loading, closing the vessel, compression, holding, and decompression and unloading is commonly referred to as the "cycle time."

Pressure come-up time (CUT): The time required to increase the pressure of the sample from atmospheric pressure to the target process pressure. The CUT depends upon the rate of compression of the sample and pressure-transmitting fluid and is proportional to the power of the pump and flow rate, size of intensifier (gpm), restrictions of high-pressure tubing and valve loss, and the target process pressure.

Decompression time or come-down time (CDT): The time to bring a food sample from process pressure to near atmospheric pressure. The CDT is almost instantaneous. The fast cooling capacity of HPP is of most interest in the production of high-quality foods.

Pressure holding time (HT): The time interval between the end of compression and the beginning of decompression. Commercial processing times, which is the sum of CUT + CDT + HT, can range from a pulse of a few seconds to over 20 min.

The studies of the effects of pressure CUT and holding time, rates of compression, and decompression on the inactivation of pathogenic organisms suspended in different medium suggested that the these HPP parameters should be taken into consideration as a factor affecting microbial inactivation and subsequent storage and carefully documented (Syed et al., 2014). Short cycle times must be targeted since process time has a significant effect on the commercial economics of HPP. A processing time less than 5 min is preferred to maximize productivity and economically justify use of the technology. Obviously, long CUTs will add appreciably to the total processing time and affect the process throughput, but these periods will also affect the inactivation kinetics of microorganisms. Therefore, consistency and control of the CUT, HT, and CDT are important in the development of HPP techniques.

4.4 PRESSURE

Process pressure (P) refers to the holding pressure during the sample treatment. Due to the system leaks, the pressure drop occurs during hold time. Maximum pressure drop should not be exceeded during HPP operation and needs to be accurately recorded. The pressure achieved and the level of reading accuracy should be identified. It has been recommended to control and record pressure at F 0.5% (electronic) or F 1.0% (dial display) level of accuracy. Further, it is recommended to have at least two methods to measure pressure and have an appropriate periodic calibration program. Frequently, a reference sensor or gauge is sequestered from routine operation and only used in a periodic check mode. Since most HPP users do not have calibration equipment in-house for high pressure, experienced laboratories should perform the calibration of any reference device

Pressure pulsing: Treatment by the application of two or more pressure pulses. When necessary the number of pulses, interval between pulses, temperature and pressure have to be identified.

4.5 TEMPERATURE

Product initial temperature (T_i): The initial temperatures of the product before the commencement of pressurization.

Product process temperature (T_f): The final temperature of the product during the sample treatment.

Process temperature: Temperature of pressure-transmitting fluid (water) during holding time.

Initial temperature of pressure-transmitting fluid and the process vessel must be documented, as they are integral conditions for microbial inactivation. Similar to a thermal process, sufficient equilibrium time (primarily based on dimensions and thermal properties of the test product sample and vessel) should be given to assure that all locations within the pressure vessel and the product are equilibrated within $\pm 0.5°C$ of the target initial temperature. For heterogeneous food samples, additional care or time may be needed to achieve equilibration. An adequate description of methods used for preheating or controlling the sample temperature before HPP should be provided.

The temperature control includes control of the temperature of the pressurization water from $5°C$ to $30°C$. However, there is no control of the water temperature inside the vessel. The high-pressure vessel does not have any temperature control, and it is at the temperature of the room where it is located.

It has been recognized that high pressures of interest do not influence the type K thermocouple readings at temperatures below $500°C$. The reference thermocouple sensor should be located at the cold point or equivalent zone within the pressure vessel and calibrated to an accuracy of F $0.5°C$ (Farkas and Hoover, 2000). Standard methods and good laboratory practices regarding temperature measurement should be followed.

It is important to consider that all compressible substances change temperature during physical compression and this is an unavoidable thermodynamic effect. In general, the temperature of both the product and the pressure-transmitting fluid may rise by 20–40°C during HPP treatment. Solid metallic materials do not experience significant compression heating. Therefore, the temperature increase may vary in foods with relatively complex compositions (Patazca et al., 2007).

In HPP sterilization, compression heating can be used advantageously to provide heating to the targeted final temperature without the presence of large thermal gradients. The compression of the product produces a consistent and thermodynamically predictable temperature increase. Using the minimum initial temperature, the final temperature is estimated for the final process. For example, for a final process temperature of 110°C at 670 MPa, an initial temperature of 78°C results in a 22°C increase from compression heating.

Preheating, precooling, and temperature equilibration of products are important steps in HPP to achieve the required process temperature. A uniform initial target temperature T_i of the food sample is desirable in order to achieve a uniform temperature increase in a homogeneous food system during compression. If cold spots are present within the food or the food system is not homogeneous, parts of the product will not achieve the target process temperature T_f during pressurization. In other cases, keeping the initial temperature low (~ 4°C) can assist in avoiding a temperature increase during pressurization, and a precooling step may be required.

Detailed information on CUT, holding time, CDT, medium pressure and temperature, product pressure, and temperature (before, during, or after processing) are usually not provided in the published studies. Sometimes the pressure increasing and/or decreasing rates are given. The effects of HPP process parameters as reported in reviewed literature are summarized in Table 4.1.

4.6 PRODUCT PARAMETERS

HPP process variables may be further complicated by the presence of variability of food composition (e.g., fat content and its distribution), pH, and water activity of the food.

4.6.1 pH
Inactivation of vegetative pathogenic microorganisms of primary public health significance and inactivation of spoilage microorganisms in

Table 4.1 Summary of Effects of Process Parameters on HPP Pasteurization				
HPP Process Parameter	Range	Effect on Lethality Reported	Facility to Control Parameter in Processing	Research Performed
Come-up time, min	2−3	Yes	No	Limited
Come-down time, min	0.3−0.4	Yes	No	
Holding time, min	3−5	Yes	Yes	Yes
Process time CUT + HT + CD, min	≥ 5	Yes	Yes	Yes
Process pressure, MPa	400−600	Yes	Yes and No	Yes
Initial temperature, °C	4−20	No	Yes	Reports of initial temperature
				No effect
Process temperature, °C	20−50	Yes	Yes	Available
Product temperature, °C	4−30	No	No	No

high acid (pH < 3.5) and medium acid (3.5 < pH < 4.6), and low acid foods (pH > 4.6) is a requirement for HPP pasteurization that is conducted at pressures above 200 MPa and at chilled or process temperatures less than 45°C. Reduced pH is generally synergistic with pressure in eliminating microorganisms. Reduced water activity, however, tends to inhibit pressure inactivation, with noticeable retardation as the water activity falls below ~0.95.

Pressurization of food constituents causes the ionic dissociation of water molecules with a corresponding decrease in pH (Hoover et al., 1989). The dissociation in aqueous solutions is influenced by temperature and pressure leading to quite different reversible pH changes of pressurized buffer systems. The effect is more pronounced for phosphate buffers than water-based food systems. On the other hand, Tris−HCl type buffers are more stable under pressure. As pH is lowered, most microbes become more susceptible to HPP inactivation, and recovery of sublethally injured cells is reduced. Accordingly, the type and composition of the buffer used in the microbial testing should be documented when HPP microbial inactivation data is analyzed.

Samaranayake et al. (2013) measured pH *in situ* and reported the results for liquid and semiliquid foods under pressure up to 800 MPa. The foods subjected to *in situ* measurements included grapefruit juice pH of 3.30; ranch dressing 3.35; apple juice 3.65; orange juice (unsweetened) 3.90; guacamole 3.90; yogurt (plain nonfat) 4.28; tomato juice 4.30; distilled deionized water 5.50; chicken broth (99% fat free) 5.80, and milk 6.60. The measurements showed an increase of acidity in all tested liquid food and water samples with increasing pressure. In general, food samples exhibit a buffering action under pressure resisting changes in pH. Unlike the pH of distilled deionized water, the pH of fruit juices and milk remained almost unchanged up to 100 MPa, and, then, gradually dropped and leveled off beyond 500 MPa. The maximum reduction of pH was, however, 0.3 pH units in the entire pressure range (0.1−800 MPa). For chicken broth, we observed a drastic pH drop (-0.37 ± 0.02 pH units) ($P \leq 0.05$) with increasing pressure up to 100 MPa, and, then, the pH further dropped an average of 0.05 pH units per 100 MPa and bottomed out at 500 MPa. Increase in pH beyond 500 MPa was, however, insignificant ($P > 0.05$), averaging about 0.03 pH units per 100 MPa. As these foods are known to contain weak acid buffering agents, the observed buffering action under HPP is therefore not surprising. The semiliquid foods also underwent a drastic pH drop (in the range of -0.3 to -0.4 pH units) ($P \leq 0.05$) at 100 MPa. For ranch dressings and yogurt, the pH again dropped significantly ($P \leq 0.05$) at 200 MPa, and, then, bottomed out between 300 and 600 MPa, whereas for guacamole, the pH was virtually unchanged for the rest of the pressure range (100−800 MPa). The maximum pH drop (-0.58 ± 0.05 pH units) was observed with yogurt in the pressure range of 300−600 MPa.

Water activity (a_w), humectants, redox potential, preservative levels, and acids used in product devilment should be measured and reported including whether or not the menstruum is a buffer, microbiological medium, or food system. Studying the effect of water activity on the HPP inactivation of *Listeria innocua*, Mariappagoudar (2007) found that reducing the water activity reduces the inactivation rate and results in an increase in *D*-values. These results appear to be consistent with previous research done by Palou et al. (1997) and Setikaite et al. (2009), who studied the combined effect of the water activity a_w and HPP processing on the inactivation of *Zygosaccharomyces bailli* and *Escherichia coli*.

Food matrix may provide additional protective or inhibitory effects. Compared to orange and tomato juices, milk showed a considerable

Table 4.2 Summary of Effects of Product Parameters on HPP Pasteurization

Product Parameter	Range Investigated	Effect on Lethality of HPP	Controlled by Product Formulation	Research Studies Published
pH	3–7	Yes	Yes	Yes
Aw	0.9–1.0	Yes	No	Yes/No
Composition	Liquid	Yes	Yes	Limited
	Semiliquids			
	Solids			
Soluble solids (Brix)	Juices	Yes	Yes	Limited
Fat content	Depends on product type	Yes	No	Limited

baroprotective effect against HPP inactivation *of Salmonella enterica* and *Listeria monocytogenes*. The inactivation patterns of *S. enterica* and *L. monocytogenes* varied in different food matrices. Decimal reduction times (*D*-value) for *S. enterica* treated 300 MPa were 0.63, 0.94, 0.41, and 0.45 min, respectively, in deionized water, milk, orange juice, and tomato juice, while those for *L. monocytogenes* were 1.40, 9.56, 1.11, and 0.94 min.

The effects of food product parameters reported in literature are summarized in Table 4.2.

CHAPTER 5

HPP Microbial Effects in Foods

The inactivation of essential groups of food microorganisms by pressure, their resistance to pressure, and factors to consider in order to develop preservation process using HPP are described in this section. The determination of inactivation rate parameters such as D- and z-values is given to characterize microbial reduction under pressure.

The efficiency of HPP to inactivate microorganisms is dependent on the target pressure, process temperature, and HT. Most of the vegetative microorganisms, yeasts, and viruses can be inactivated at or near room temperatures. It is worth noting that pressure resistance of microorganisms is often at its maximum at room temperature. The resistance decreases either by processing at chilled or moderately heated conditions (Buckow and Heinz, 2008). On the other hand, bacterial spores are highly pressure resistant, and combination of pressure (400–600 MPa) and heat (90–120°C) are required for sterilization of most molds. Extensive reviews on HPP are available documenting the impact of pressure treatment on food safety and quality (Farkas and Hoover, 2000, Hendrickx et al., 2002).

Inactivation of vegetative pathogenic microorganisms of primary public health significance and inactivation of spoilage microorganisms in high acid (pH < 3.7) and medium acid (3.7 < pH < 4.6) foods is a requirement for HPP pasteurization that is conducted at pressures above 200 MPa and at chilled or process temperatures less than 45°C. Pasteurization demands a logarithmic reduction of 5 or 6 in pathogens. *Listeria monocytogenes* is considered as a target microorganism of public health concern in dairy and meat products, *Salmonella* is a target microorganism in eggs, and *Escherichia coli* is a target microorganism in fruit and vegetable juices and other fruit-based products.

Low acid juices are inclusive of fruit and vegetable preparations with a pH > 4.6 and water activity >0.85. Juices prepared from low acid fruit and vegetables represent a significant food safety issue given the potential to support pathogen growth including spore formers such as *Clostridium botulunium*. Indeed, an outbreak linked to thermally

Adapting High Hydrostatic Pressure for Food Processing Operations. DOI: http://dx.doi.org/10.1016/B978-0-12-420091-3.00005-4

pasteurized carrot juice in 2006 resulted in six confirmed cases of botulism.

The rate of inactivation is strongly influenced by the peak pressure (Patterson, 2005). Commercially, higher pressures are preferred as a means of accelerating the inactivation process, and current practice is to operate at 600 MPa, except for those products where protein denaturation needs to be avoided. The pressure resistance of vegetative microorganisms often reaches a maximum at ambient temperatures, so the initial temperature of the food prior to HPP treatment can be reduced or elevated to improve inactivation at the processing temperature (i.e., the temperature at pressure). The extent of inactivation also depends on the type of microorganism and on the composition, pH, and water activity of the food. Gram-positive bacteria are more resistant than Gram-negative. Most yeast is inactivated by exposure to 300–400 MPa at 25°C within a few minutes. However, yeast ascospores may require treatment at higher pressures. The pressure inactivation of molds appears to be similar to yeast inactivation. However, additional studies on molds of interest in food preservation are needed. Reduced pH is generally synergistic with pressure in eliminating microorganisms. Reduced water activity, however, tends to inhibit pressure inactivation, with noticeable retardation as the water activity falls below ~0.95.

5.1 DEVELOPMENT OF HPP PRESERVATION PROCESS

The establishment of processing conditions for HPP pasteurization and sterilization processes requires optimization of the process temperature and pressure to inactivate the target pathogenic and spoilage-causing bacteria, based on knowledge of the behavior of the food under pressure and of the rate constants for microbial inactivation.

In the development of a "preservation specification" for a thermal pasteurization or sterilization process, the processing time F_p is traditionally defined by the initial load of resistant organisms N_0, the endpoint of the process N_F, specific logarithmic reduction or SLR, and the logarithmic resistance D of the target bacteria (D-value) under defined conditions, and it is calculated using Equation (5.1):

$$F_p = D \times (\log N_0 - \log N_F) = SLR \times D \tag{5.1}$$

A kinetic analysis of microbial destruction is carried out under HPP isobaric and isothermal conditions after the CUT portion of the pressure cycle has been completed and the maximum process pressure attained. The D-value can be obtained as the negative reciprocal slope of $\log(N/N_0)$ vs. time t and is reciprocally related to k. The thermal D-value at constant pressure, $D_{T,P}$, can be calculated from survival curves plotted at constant pressure, as expressed by the following equation:

$$D_{T,P} = \frac{t}{\log(N/N_0)} \tag{5.2}$$

$$D_{P,T} = \frac{t}{\log(N/N_0)} \tag{5.3}$$

The thermal resistance z_T (the temperature required to reduce the value of $D_{T,P}$ by 90% at constant pressure) and the pressure resistance z_P (the pressure required to reduce the value of $D_{P,T}$ by 90% at constant temperature) are determined from the following equations:

$$z_T = -\frac{1}{[\text{slope}]} = \frac{T_2 - T_1}{\log(D_{T_1,P}/D_{T_2,P})} \tag{5.4}$$

$$z_P = -\frac{1}{[\text{slope}]} = \frac{(P_2 - P_1)}{\log(D_{T,P_1}/D_{T,P_2})} \tag{5.5}$$

Table 5.1 presents the examples of generated D- and z-values for the most significant pathogenic organisms of concerns and spoilage organisms.

However, a phenomenon that is common to HPP is the persistence of pressure-resistant microorganisms during treatment. When microorganisms are subjected to high pressure, initially there is a linear decrease in the microbial load; however, as time progresses, the rate of microbial inactivation decreases, ultimately resulting in a survival curve with a "pressure-resistant tail." The existence of this subpopulation of pressure-resistant microorganisms is poorly understood. The "tailing effect" may be due to inherent phenotype variation in pressure resistance among the target microorganisms; however, the substrate and growth conditions may also have an influence. Biphasic kinetics for the destruction of vegetative cells has been applied to evaluate HPP inactivation: the first phase is a kill due to the pressure pulse, and a first-order kinetic model is applied to the microbial kill during the HT.

Table 5.1 Kinetic Parameters for Inactivation of Microbial Population by HPP in Different Food Matrixes

Microorganism	Substrate	Time Parameter		Pressure $z(T)$ and $z(P)$ (MPa)	Other	References
		D	k			
Units		min	1/min			
		Vegetative Cells				
Campylobacter		<2.5	>0.92	300		Smelt and Hellemons (1998)
Salmonella enteritidis	Meat	3	0.768	450		Patterson et al. (1995)
Salmonella typhimurium	Milk	3	0.768	350		Patterson et al. (1995)
Yersinia enterocolitica	Milk	3	0.768	275		Patterson et al. (1995)
E. coli	Milk	1	2.303	400	$T=50°C$	Gervilla et al. (1997a)
E. coli	Meat	2.5	0.92	400		Patterson and Kilpatrick (1998)
E. coli	Milk	1	2.303	450	$T=25°C$	Gervilla et al. (1997a)
E. coli O157:H7	Milk	3	0.768	400	$T=50°C$	Patterson and Kilpatrick (1998)
Staphylococcus aureus	Milk	2.5	0.92	500	$T=50°C$	Patterson and Kilpatrick (1998)
S. aureus	Meat	3	0.768	500	$T=50°C$	Patterson and Kilpatrick (1998)
L. monocytogenes		1.48–13.3	0.173–1.556	350	101 strains	Smelt and Hellemons (1998)
L. monocytogenes	Milk	3	0.768	375		Patterson et al. (1995)
L. monocytogenes	Meat	2.17	1.061	414	$T=25°C$	Ananth et al. (1998)
L. monocytogenes Scott A	Meat	3.5	0.658	400	$T=$ ambient	Mussa et al. (1999)
Listeria innocua	Eggs	3	0.768	450	$T=20°C$	Ponce et al. (1998)
L. monocytogenes	Ground Pork	1.89–4.17	0.552–1.219	414	$T=25°C$	Murano et al. (1999)
L. monocytogenes	Ground Pork	0.37–0.63	3.656–6.224	414	$T=50°C$	Murano et al. (1999)

Spores

Organism	Substrate						Reference
Clostridium sporogenes		16.772	0.138		600	T = 90°C	Rovere et al. (1996a)
C. sporogenes		6.756	0.341	725 (90°C)	700	T = 93°C	
C. sporogenes		5.306	0.434		800	T = 93°C	
C. sporogenes		1.282	1.796		600	T = 108°C	Rovere et al. (1996a)
C. sporogenes		0.901	2.556	752 (108°C)	700	T = 108°C	
C. sporogenes		0.695	3.314		800	T = 108°C	
Clostridium botulinum type E Alaska	Buffer	8.77	0.263		758	T = 35°C	Reddy et al. (1999)
C. botulinum Type E Beluga	Crab meat	3.38	0.681		758	T = 35°C	Reddy et al. (1999)
C. botulinum Type E Beluga	Crab meat	1.64	1.404		827	T = 35°C	

Yeast

Organism	Substrate						Reference
Saccharomyces cerevisiae	Orange juice	10.81	0.21		300	T = 34°C	Zook et al. (1999)
		2.8	0.82		350	T = 36.8°C	
		0.97	2.37	117	400	T = 37.2°C	
		0.5	4.61		450	T = 39.7°C	
		0.18	12.79		500	T = 43.4°C	
S. cerevisiae	Apple juice	9.97	0.231		300	T = 34°C	Zook et al. (1999)
		0.88	2.617	115	400	T = 37.2°C	
		0.28	4.798		450	T = 39.7°C	
		0.15	15.35		500	T = 43.4°C	
S. cerevisiae		1.27	1.813		350	pH = 3.7	Parish et al. (1998)
S. cerevisiae		0.067	34.373		500	pH = 3.7	

Source: Adapted from Kinetics of Microbial Inactivation for Alternative Food Processing Technologies: http://www.fda.gov/Food/FoodScienceResearch/SafePracticesforFoodProcesses/ucm100198.htm

In the case of survival curves showing shoulders or tailing that deviate from a linear curve, Weibull distribution model given by the following equation can be applied:

$$\log \frac{N_t}{N_0} = - bt^n \qquad (5.6)$$

where b is a scale factor and n is a shape parameter: $n < 1$ for curves that are concave upward, $n > 1$ for curves that are concave downward and $n = 1$ for linear curves. The scale factor is considered to be a non-linear rate parameter, which primarily reflects the overall steepness of the survival curves when the power n is fixed.

CHAPTER 6

Emerged HPP Commercial Applications

Application of HPP for raw and RTE meat and poultry products, fruits and vegetable processing, cold pressed juices, seafood and dairy categories are discussed. The range of operating pressures to be applied to achieve shelf-life extension, pasteurization, and nonthermal extraction treatments of different categories of finished and raw seafood products are presented.

In the early 1990s, the first commercial application of high-pressure technology to food was observed, with the application of HPP treatment to a high acid jam produced by the Japanese company Medi-Ya. More recently, the use of high pressure on foods has extended to meat-based products (31% of all industrial applications), vegetable products (35%), juices and beverages (12%), seafood and fish (14%), and other products (8%). Varieties of products processed by HPP are commercially available in North America, Europe, Oceania, and Asia.

6.1 MEAT PROCESSING

There are three key areas where meat processors use HPP: (1) chemicals free preservation with extended shelf-life and enhanced food safety while maintaining the appearance, flavor, or consistency that consumers want and need; (2) development of new products with enhanced functionality, and (3) harvest processing (Figure 6.1). Due to the differences in pressure effects at various pressurization levels and nature of meat products, development and optimization of pressure–temperature processing conditions is required for these operations.

In order to inactivate bacterial spores during HPP-assisted sterilization, a combination of pressure at levels higher than 700 MPa and elevated initial temperature ($>80°C$) seems to be a promising improvement to the traditional heat sterilization of low acid foods. Combining pressure and heat treatments can result in a sterilization of food product at reduced thermal load and consequently result in higher

Adapting High Hydrostatic Pressure for Food Processing Operations. DOI: http://dx.doi.org/10.1016/B978-0-12-420091-3.00006-6

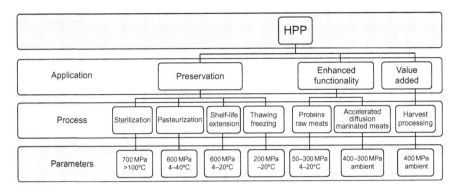

Figure 6.1 HPP applications in meat processing.

nutritional quality. The US FDA was enabled to certify a PATS process of low acid foods in 2009.

HPP is mainly used in a commercial environment as an effective postlethality technology for prepacked meat products. The majority of the HPP commercial meat products belong to RTE food category. In the United States, Hormel Foods is using HPP mainly for the elimination of *L. monocytogenes*, since the United States Department of Agriculture (USDA) regulatory authority maintains a policy of zero tolerance. Pressure levels applied for the pasteurization of meats and meat products, range in an area of 400–600 MPa with short processing times of 3–7 min at ambient temperature. In most cases, these treatments lead to 5-log reduction for the most common vegetative pathogenic (*E. coli*, *Listeria*, and *Salmonella*) and spoilage microorganisms resulting in an increased shelf-life and improved safety. The Spanish meat processor Espuña uses HPP equipment for decontamination of cured, cooked ham with the aim of prolonging the shelf-life to 60 days in an undisturbed cooling chain.

In addition to extending shelf-life and increasing food safety when used as a processing step, HPP can enable innovation through the creation of ingredients and finished products with unique functional properties. For example, HPP can be used to blanch meat products, a common process that typically uses steam or hot water to soften or partially cook some foods. Blanching is used to reduce cooking time—a key attribute in convenience food products. Avoiding the use of hot water, HPP accomplishes this without heat damage to the product which protects its color and taste.

In the case of fresh meat treatment, HPP can affect quality parameters particularly depending on the pressure level applied, and thus typical characteristic associated with fresh meat like texture and especially color can be remarkably modified. Additionally, the application of HPP for fresh meats leads to different degrees of protein structure modification. As a general mechanism, the application of pressure induces unfolding of the protein structure and subsequent folding after hydrostatic pressure release and leads depending on the specific protein and conditions applied to partial or total denaturation. The application of low HPP levels up to 300 MPa can be used to improve the functional and rheological properties of poultry meat, tenderize prerigor meat, and to improve the water retention properties of raw material in the development of products with reduced salt content such as sausages.

So far only one fresh meat product, minced beef, which is further processed before its consumption is available in the pressurized form in the market. Cargill (NE, USA) is the meat producer known to be using high-pressure pasteurization on raw beef. Beef is made into ground beef patties then packaged and shipped to American Pasteurization Co. in Milwaukee for HPP. After vacuum packaged, Fressure beef patties undergo HPP; they are packaged and shipped to food distributors like Sysco.

Hormel (MN, USA) applies HPP to the animals immediately after harvesting to reduce fecal contamination, to remove hair and feather, and to loosen nails from the hoof. Immediately after the animal is harvested, it is subjected to HPP to stop glycolysis early and subsequently to drop the pH, which affects tenderization, water holding capacity, and the overall likeability. The process at 200 MPa keeps the pH around 6.2–6.4, which is unachievable by any other means.

6.2 POULTRY

Purdue Farms, Oscar Mayer, Tyson Foods, and Hormel are big-name companies who are utilizing high pressure to add value to poultry offerings. Maple Lodge Farms (Ontario, Canada) was the first poultry processor in Canada to incorporate HPP technology into its operations. Maple Lodge has implemented a new method of ensuring that precooked sliced luncheon meat that has already been packaged is safer for consumers. In addition to the added safety benefits, the process also reduces the need for the company to add preservatives and other chemicals to its products, making it possible to make healthier and more natural food.

6.3 FRUIT, VEGETABLES, AND JUICE PRODUCTS

HPP was introduced to consumers with the taste of all-natural refrigerated guacamole (Fresherized Foods, USA), soon followed by ripe avocado halves with a shelf-life of up to 30 days. Later, the use of HPP technology in fruit and vegetable processing expanded to salsa, pre-chopped onions, cold pressed organic juices, fruit smoothies, and apple sauce (Leahy Orchard Inc., Canada). HPP inactivates *Salmonella, E. coli,* and *L. monocytogenes* in fruit and vegetable liquid and semiliquid products. The color of fruit and vegetable products such as jams, fruit juices, and purees is generally preserved if thresholds of temperature and/or pH are observed. Excellent retention of fresh-like flavors for far greater time periods than those obtained with conventional thermal treatment has also been observed under optimal storage conditions.

HPP offers safety and shelf-life extension advantages for juices and other beverages—without adding heat or preservatives. HPP has gained acceptance as a safety step in beverage processing and its use is seeing rapid growth, fueled by market factors such as the high demand for safe and nutritious fruit and vegetable juices, coconut waters, and fruit purees. New so-called clod pressed detox beverages have been launched during last 5 years. HPP between 400 and 600 MPa from 1 to 10 min reduces several log reduction of spoilage organisms and pathogens such as *E. coli* O157:H7, *Salmonella, Listeria,* and *Cryptosporidium.* Some of the advantages of HPP for processing fresh juices include development of premium healthier juices with a higher level of functionality, avoid postcontamination and increase safety, expand markets, and protect the brand. Processors use HPP in their HACCP program to achieve the US FDA requirement of a 5-log reduction in pathogens in fresh juices.

HPP cold pressed juices from Evolution Fresh, BluePrint and Suja (USA) dominate the super-premium juices market in the United States. FreshBev LLC recently registered "craft juices" and "craft bar juices"—a new line of "raw organic cleanse" that are made of freshly extracted juices are bottled within the hour, HPP processed, and sold next day.

European companies presently employing this technology include UltiFruit® (for orange juice), the Pernod Ricard Company, France, and Solofruita, Italy (for fruit jams), Fruity Line, The Netherlands, UGO, Slovakia. ColdPress was launched in the United Kingdom in 2011, Preshafood is located in Australia, and COL + is in New Zealand.

6.4 DAIRY PRODUCTS

In the dairy industry, the potential for the use of HPP technology exists not only from the standpoint of microbial and enzyme inactivation in raw milk but also for the purpose of improving the quality and yield of dairy products such as cheese, yogurt, and exploring new functional properties of traditional dairy ingredients such as casein proteins. The selective barosensibility of spoilage-causing and pathogenic microorganisms has been used in the postpackaging HPP of cultured foods, as was reported by Hyperbaric Company, Spain. Inactivation of yeast and molds resulted in up to 3 months' conservation of a reduction in the *Lactobacillus* count and retention of bioactive components such as lactoferrin and inmunoglobulins, without alteration of their physiological properties. In addition, HPP is being used for the selective control of yogurt starter cultures.

6.5 SEAFOOD AND FISH

Clean, virtually 100% separation of meat from lobsters, oysters, clams, and other fresh products can be achieved by pressure denaturing of the specific protein that holds the meat to the shell. HPP allows the maximum product yield without causing any mechanical damage to the product, regardless of size. By adjusting the processing conditions, beneficial texture changes can also be created by improving the moisture retention ability of proteins, thus resulting in less water loss during storage or cooking. Additionally, HPP is being used as a means to accomplish nonthermal treatment for reducing the bacterial load of raw or fresh seafood products. Oysters are a natural match for HPP, as are fresh juices, sashimi, and any product where disease outbreaks have been associated with minimal processing of the product. HPP reduces many of the risks associated with the consumption of raw oysters. It has been shown that many *Vibrio* (*parahaemolyticus*, *cholerae, vulnificus*) microorganisms are destroyed by HPP operating pressures.

HPP can also cause the inactivation of certain enzymes that result in the deterioration of fish products. While the decomposition of proteins and lipids are primarily due to enzymes resulting from microbial contamination, there are many fish and shellfish species that have active enzyme systems of their own, that also contribute to product spoilage at refrigeration temperatures. The inactivation of microorganisms and enzyme systems in seafood products has economic and food safety significance.

Seafood pates and salads are ideal products for HPP treatment. The further processing required to produce these latter products provides an opportunity for control of microbial contamination with spoilage and pathogenic microorganisms. Also, the incorporation of raw vegetables (such as parsley, cilantro, green onions, garlic) in salads can result in high mold populations exceeding the product specifications established by many institutional buyers.

According to Hiperbaric, the pressure levels for the elimination of spoilage microorganisms and shelf-life increase in seafood and fish are in the range of 400−600 MPa from 1 to 5 min. For nonthermal shucking and meat extraction pressures between 200 and 350 MPa are able to shuck bivalves and to achieve yield increase by 20−50% and reduce hand labor.

CHAPTER 7

Emerging Applications

The chapter presents new processing opportunities when HPP is combined with low temperatures in freezing and thawing operations for premium meat and fish products. Additionally, HPP potential in developing meat products with reduced sodium chloride content is discussed.

In addition to reducing microbial levels by inactivation, HPP can offer new opportunities for innovation and new products development by inhibiting and stabilizing microbial growth during subzero storage with and without freezing. Freezing of meat provides a safe and convenient way of shelf-life extension without negative effects on the nutritional quality. When elevated pressure combined with temperatures in low subzero diapason, it can directly affect the phase or state of water that can be controlled. Numerous interesting effects of high pressure on the solid—liquid and solid—solid phase transitions of water can be achieved taking advantage of phase diagram and manipulating of pressure and temperature.

7.1 HPP AT LOW TEMPERATURES

Increased hydrostatic pressure influences the phase transition of water by way of depressing the freezing/melting point to a minimum of $-22°C$ at 210 MPa. Besides a depression of the freezing point, a reduced enthalpy of crystallization can be observed, thereby accelerating phase transition processes. Furthermore, different solid states of pure water with a higher density exist under pressure above 210 MPa known as ice I—V, but these solid forms are stable only under high pressure.

Taking advantage of the phase diagram of water shown in Figure 7.1, various pathways of changing the physical state of food can be followed using external manipulations of temperature and/or pressure. At atmospheric pressure freezing crystallization happens when temperature falls below 0°C (A—I—E or F). Definitions on possible HP—LT processes can be given based on a terminology introduced by Knorr et al. (1998). The processing steps illustrated on the water

Adapting High Hydrostatic Pressure for Food Processing Operations. DOI: http://dx.doi.org/10.1016/B978-0-12-420091-3.00007-8

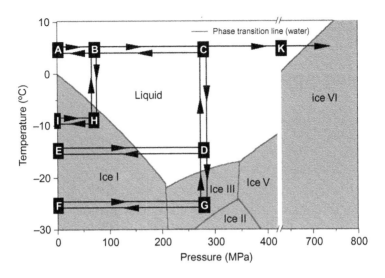

Figure 7.1 Phase diagram of water.

phase diagram can range from storing food under high pressure at sub-zero temperatures without freezing (A−B−C−D−C−B−A) to freezing at temperatures above 0°C (A−B−C−K−ice VI). According to water phase diagram, a food sample may be pressurized, e.g., to 112 or 207 MPa then cooled down under pressure to about −10°C or −21°C respectively, without ice crystal formation. It may be useful to store some biological samples under such conditions to prevent microbial growth, slow down most deterioration reactions, and avoid freezing damage.

Some of the more interesting possibilities are pressure-assisted freezing (A−B−H−I, that includes pressurization of an unfrozen sample, cooling, freezing at nearly constant pressure, pressure release) and pressure-assisted thawing (I−H−B−A, i.e., pressurization of a frozen sample, heating, thawing at nearly constant pressure, warming, pressure release).

There is also a process of pressure-shift freezing (A−B−C−D−E) where crystallization is induced simultaneously in the whole sub-cooled sample by fast pressure release, with the intention of obtaining small and uniform ice crystals in the sample with minimum damage to the tissue. The reverse of the above process is pressure-induced thawing (E−D−C−B−A) where the phase change is induced by pressurization.

For freezing or thawing under pressure, food samples must be protected by packaging material from contact with the cooling and pressurizing medium. The packaging should possess flexibility to transmit pressure, have barrier properties against liquids, and resistance to freezing, piercing, and delamination.

7.2 POTENTIAL IMPLEMENTATIONS OF LOW-TEMPERATURE HPP FOR MEAT PRESERVATION

The advantages and economic feasibility of subzero preservation of food products under pressure can be possible to demonstrate by developing high-value products with improved texture, consistency, and smoothness using fast freezing/thawing rate. Freezing rate is important process parameter defined by product-specific qualities that affect product quality. Slow freezing results in larger ice crystals, which generally damage the texture of the food, whereas a rapid freezing rate usually preserves food texture. Rapid pressure-assisted freezing can be achieved by cooling at 20°C and 209 MPa when water remains in the liquid state. Upon release of pressure, instantaneous and homogeneous crystallization occurs with formation of very small crystals. Research conducted with small pieces of meat with various shapes and geometry showed that pressure-assisted freezing of pork muscles could provide a uniform small-size ice crystal distribution in comparison with air-blast freezing and liquid nitrogen freezing at atmospheric pressure.

Pressure-assisted thawing appears to be of special interest because it can be carried out mostly at subzero temperatures, thus reducing microbial hazards and some other disadvantages of thawing near 0°C, with a much shorter processing time and usually with less exudation phenomena. Pressure thawing of ground beef with diameters ranging from 55 to 80 mm has been be efficiently achieved at 210−280 MPa for 30 min and with no quality loss compared to thawing at atmospheric pressure (12 h at 3°C). Additionally it was found that the transient phase change of ice I to ice III during pressurization of frozen systems appear to be an effective way to reduce bacterial contamination. Pressure-shift freezing can offer higher meats quality in terms of lesser drip loss and unchanged firmness after thawing and should be further investigated on larger food samples. Concept exploring and feasibility research was done using lab or pilot scale units with meat, seafood, and fruit and vegetable products. HP−LT processes are not industrially implemented. Pressure and temperature conditions should be carefully chosen and scale-up to an industrial level will require accurate control of the temperature profile in large-scale devices. The potential benefits of low-temperature HPP can be investigated in the case of meat offal, RTE meals (high-value products containing meat and/or fish), and marine products such as oysters, scallops, fish eggs, and shrimps.

7.3 HPP FOR SODIUM REDUCTION IN RTE MEATS

Reducing the salt levels in meat products represents a significant challenge due to the sensory and functional roles salt has in processed meat products. Meat processors have attempted to meet consumer demand for low salt products by using KCl as an alternative to NaCl. It is possible to substitute NaCl with KCl because the main role of salt is primarily due to the action of chloride ions that cause swelling of the myofilaments. Although high K + ion levels can lead to renal disease especially in members of the population suffering with Type I diabetes. Therefore, reduction in salt is preferred by finding processing alternatives. A more significant role of salt in meat products relates to changes induced in protein functionality that enhances water holding capacity, in addition to contributing to desirable texture characteristics. Salt added to processed meats results in the solubilization of the functional myofibrillar proteins in meat that subsequently adds to the binding properties of meat, hence texture. Increasing the water holding capacity of the meat reduces cook loss thereby increasing tenderness and juiciness of the meat product. It follows that reducing the salt content of meat products leads to decrease swelling of mycofibers with the net result of poor water holding capacity and excessive exudate, in addition to loss of texture.

Although widely studied, HPP has not been successfully applied to fresh meats as a commercial treatment. As discussed by Bajovic et al. (2012), one of the major problems with fresh meat is that HPP-treated meat looks cooked to the consumer because of the pressure-sensitive protein denaturation. However, according to Ma and Ledward (2013), if prerigor meat is subjected to pressures of about 100–150 MPa, below those necessary to cause color changes, it becomes significantly more tender than its untreated counterpart and this may now be a commercially viable process, given the decreasing cost of high-pressure rigs. When treated at 100–200 MPa while the temperature is raised from ambient to around 60°C, postrigor meat also yields a tender product and this may also be a commercially attractive process to parts of the food industry, e.g., those involved in catering.

Another study into using HPP in place of salt was reported by Crehan et al. (2000) who reported on the sensory and water holding capacity of frankfurters. The researchers treated the raw batter mix containing 1.5% NaCl with pressure treatments up to 300 MPa. The conclusion from the study was that HPP at the highest level applied resulted in comparable water holding capacity and texture characteristics as formulations prepared using 2.5% salt. In the study aimed to determine the interactions between salt (NaCl), sodium phosphate (SP), and mild HPP in brine injected beef, Ma and Ledward (2013) have shown that the HPP increased the effectiveness of salt and phosphates on protein functionality.

Sufficient research has not been conducted on innovative processing or novel technologies that potentially can reduce NaCl or to substitute low salt, low sugar, and low fat in the processed foods. Research on investigation of HPP opportunities for sodium reduction in meats is under way in Guelph Food Research Center of AAFC, Alberta, Oregon State University, and under Pleasure research umbrella in France. The focus is made on optimizing of HPP process (pressure-process temperature-hold time) along with developing of the process supported formulation that will result in chemicals and preservatives reduction and consequently safe products.

CHAPTER 8

HPP Commercial and Pilot Equipment

Mode of operation, designs of pressure vessels, and the manufacturers of commercial and pilot HPP equipment are reviewed in this chapter. The range of available and new commercial HPP machines for different groups of food producers is presented. The process economics, energy requirements along with the considerations on how to select the HPP unit are discussed.

8.1 MODE OF OPERATION

The HPP operation starts with loading prepacked food products in perforated baskets as shown in Figure 8.1 and the baskets are transported into the high-pressure vessel. The vessel is sealed and then filled with a pressure-transmitting fluid (normally water) and pressurized by the use of a high-pressure pump (an intensifier), which injects additional quantities of fluid. The packages of food, surrounded by the pressure-transmitting fluid, are subjected to the same pressure as that which exists in the vessel itself. After the product has been held for the desired process time at the target pressure, releasing the pressure-transmitting fluid decompresses the vessel. For most applications, products are held for 3–5 min at 600 MPa that allows approximately 6–8 cycles per hour including time for compression, holding, decompression, loading, and unloading. Slightly higher cycle rates may be possible using fully automated loading and unloading systems and more powerful pumps and higher throughput intensifiers and/or more pumps and intensifiers. After HPP treatment, the processed product is removed from the vessel and stored or distributed in the conventional manner.

8.2 DESIGN

The primary components of an HPP system include a pressure vessel (or shell), a closure or closures for sealing the vessel, a device for holding the closure(s) in place while the vessel is under pressure (e.g., a yoke), low-pressure fill systems to fill the pressure chamber with a temperature-regulated pressure medium (water) and remove air when

Adapting High Hydrostatic Pressure for Food Processing Operations. DOI: http://dx.doi.org/10.1016/B978-0-12-420091-3.00008-X

Figure 8.1 Perforated loading basket.

the system is closed, a high-pressure-intensifier pump or pumps to build up the pressure using a food-grade hydraulic oil, a system for controlling and monitoring the pressure and (optionally) temperature, and a product-handling system for transferring product to and from the pressure vessel. The machinery required is complex and requires extremely high precision in its construction, use, and maintenance.

8.3 PRESSURE VESSELS

Autofrettage vessels, wire-wound vessels, and heat-shrink vessels are most frequently built for systems with pressures higher than 400 MPa.

Autofrettage (French, "self-shrinking" or auto-shrinking) is a special mechanical process resulting in the pressure-vessel prestressed condition after the autofrettage process completion. The process begins by pressurizing the shell that is stopped when the mechanical stress distribution through the shell wall is such that the internal wall is deformed plastically and the external wall is deformed elastically. After the pressure is released, the elastically deformed outside shell has the tendency to regain its original shape but is thereby hindered by the plastically deformed inside part. This creates a condition when the outer part of the shell pushes continuously onto the inside part resulting in the shell permanently prestressed condition. When a shell is constructed using autofrettage, it can resist much higher internal pressures, comparing to the identical size shell made of a simple mono-block design. Autofrettage also makes shell much more resistant to crack growth.

Heat-shrink technique is developed to induce beneficial residual hoop stresses in the cylindrical shell. In heat shrink, the shell needs at least two separate layers; however, it can be made of three or more separate layers, assembled together. In the food-processing industry typically the inner layer is made of stainless steel; high strength steel alloy inner layer and high strength steel alloy outer layer. Shrink fitting is accomplished by heating up the outer layer and after achieving the thermal expansion the cool layer (with earlier inserted stainless steel liner) is assembled. Finally the whole assembly is allowed to cool down to room temperature, causing residual hoop stresses. The use of preexisting residual stresses in the cylinders before operation is essential to allow very high pressures by enhancing the load carrying capacity and increasing the number of cycles of the operation. Heat shrink increases pressure-vessel lifetime, durability, maximum pressure, and reduce the weight of shell, researchers succeeded in designing multilayer heat-shrink fitted cylinders.

Wire-winding technique combines a single layer cylinder and continuous length of wire wound around it, at the prestressed condition. It is usually applied to increase strength to weight ratio and to improve the fatigue life. The required wire-winding force is obtained based on constant effective stress in all layers at the maximum applied force. However, the failure of the wire (breaking or corrosion) will cause the pressure-vessel failure. The principle used in the wire-wound vessels is that they will leak before they break.

In some cases, the cylinders and the vessel frame are prestressed by winding wire under tension layer upon layer. The tension of the wire compresses the cylinders, reducing the diameter of the cylinders. This special arrangement allows an equipment lifetime of over 100,000 cycles at pressures of 680 MPa or higher.

Stainless steel construction is used in key areas of the vessel for reliability and easy washing down in the most demanding environments. The corrosion protection meets regulatory requirements for food-processing plants. The systems also have provisions for filtering and reusing the compression fluid (usually water or a food-grade solution). As pressure vessels of all types utilize potentially hazardous energy, the relevant regulations (ASME Boiler and Pressure Vessel Code, Section VIII Division 3) seek to identify good design, good manufacturing practices, and detailed safety assessments for the safe operation and maintenance of the vessels and auxiliary parts.

8.4 EVALUATION OF PRESENT MANUFACTURES

8.4.1 Commercial HPP Vessels

In 2013, the HPP equipment market was estimated at $350 million and is projected to grow at 11% CAGR for the next 5 years. Batch and semicontinuous systems are currently commercially available as options for HPP equipment. Batch processing that requires in-container prepackaged products is the more conventional of the two options and was relatively easy to implement in the food industry. In batch HPP systems, the product is generally treated in its final primary package; commonly, the food and its package are treated together, and so the entire pack remains a "secure unit" until the consumer opens it. In-container processing requires packages in the form of pouches, large bulk bags, or container–lid combinations. Semicontinuous HPP systems are used in only a few cases, to directly process pumpable products that then need to be aseptically packaged; 90% of HPP-processed foods are expected to be processed in flexible or partially rigid packages.

Commercial batch vessels have internal volumes ranging from 35 to 687 L. This enables the food manufacturers of all sizes to implement HPP in their operations. Avure Technologies (USA–Sweden), Hyperbaric (Spain–USA), MULTIVAC (USA–Germany), and BaoTau KeFa (China) are major suppliers of commercial-scale pressure equipment. Both horizontal and vertical pressure-vessel configurations are available. The horizontal-cylinder design is intended to serve in installations where vertical space is limited and minimal plant alteration is desired. The horizontal orientation allows single-direction inline product flow. In addition, there are differences in capacities, cycle times, achievable pressures, capital costs, etc.

HPP equipment owners also offer toll HPP services to food industries. This business model allows any food or beverage manufacturer to have access to industrial HPP equipment without the need of directly investing capital in the purchase of HPP unit. Instead, food manufactures pay for the use of existing HPP industrial plants for their service on a toll basis.

Avure Technologies (http://avure.com) offers vertically and horizontally oriented commercial pressure vessels. The typical HPP system blocks include pressure vessel, basket load conveyor, operator's panel, control system, water system, and high-pressure pumping systems as shown in Figure 8.2. Avure patented QUINTUS® wire-winding technology to contain isostatic pressure in the system. Pressure vessels and frames are prestressed and wound completely with high tensile cold-rolled spring steel ribbon. Prestressing causes the forged steel vessel wall to remain in residual compression, even when subjected to maximum loads. This result in extremely high resistance to fatigue, eliminates tensile loads, and prevents crack propagation and brittle failure. Another feature is that wire-wound vessels will leak before they burst.

The technical characteristics of industrial HPP systems built by Avure Technologies are summarized in Tables 8.1 and 8.2.

The Avure QFP 525L-600 (Figure X) measures up to the production needs of today's high-volume food processor, offering the highest throughput in the industry with features to minimize operating costs and maximize uptime. Improved filling and pressurizing efficiencies result in higher throughput, larger volume, and 10 cycles per hour with

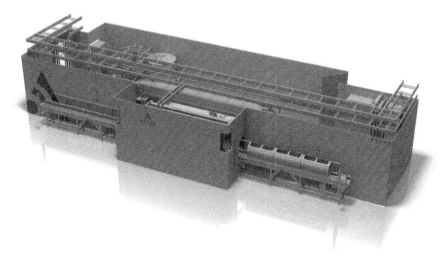

Figure 8.2 The Avure QFP 525L-600 HP unit.

Table 8.1 Vertical HPP Systems from Avure			
System Type (QUINTUS)	QFP 35L-600	QFP 215L-600	QFP 320L-400
Vessel volume, L	35	215	320
Maximum pressure, MPa	600	600	400
Maximum vessel temperature, °C	50	50	35
Cycle time, min (HT excluded)	5	4.5	3.5
Production rate, kg/cycle	25	150	2300 kg/h
Maximum filling capacity	70%	70%	
Input water temperature, °C	4–16	4–16	4–16
Weight, tons	8	37	36
Systems dimensions, m	3.5 × 3.4 × 3.5	1.2 × 0.8 × 1.6	6.7 × 8.7 × 2.7

Table 8.2 Horizontal HPP Systems from Avure				
System Type (QUINTUS)	QFP 100L-600	QFP 350L-600	QFP 687L-310	QFP 525L-600
Vessel volume, L	100	350	687	525
Maximum pressure, MPa	600	600	310	600
Maximum vessel temperature, °C	50	50	50	
Cycle time, min (HT excluded)	3	4.5	3.7	3
Production rate	70 kg/cycle	250 kg/cycle	500 kg/cycle	3690 kg/h
Maximum filling capacity	70%	70%		
Input water temperature, °C	4–25	4–16	4–16	4–16
Weight, tons	17.5	46	46	
Systems dimensions, m	6.8 × 2.8 × 1.9	6.8 × 3.9 × 2.1	6.9 × 3.9 × 2.1	

a 3 min hold time. The larger diameter also accommodates processing of larger and odd-shaped product.

The largest 687 L system from Avure, with a horizontal design, was built to serve the seafood industry, where continuous flow is of the essence. With its large 687 L capacity and 310 MPa pressure capability, this system can reliably process up to 5000 kg of raw product per hour.

Hiperbaric (Spain–USA, http://www.hiperbaric.com/en/) designs, manufactures, and markets HPP equipment since 1999. Hiperbaric HPP single vessels volume ranges from 55 to 525 L. The technical characteristics of Hiperbaric equipment type 55–420 are summarized

Table 8.3 Horizontal HPP Systems from Hiperbaric						
System Type Hiperbaric	55	120	135	300	400	525
Vessel volume, L	55	120	135	300	420	525
Maximum pressure, MPa	600	600	600	600	600	600
Number of intensifiers	1	2	2/4	4/6/8	6/8	
Cycle time, min (including HT of 3 min, loading and unloading)	6.5	6.9	7.2/5.7	7.6/6.5/ 5.9	7.7/7.0	
Production rate, kg/h	255	525	619	1303/ 1521/ 1660	1948/ 2160	3000
Maximum filling capacity	50%	50%	55%			
Systems dimensions, m	0.2 × 2.0	0.2 × 4.05	0.3 × 2.2	0.3 × 4.5	0.38 × 4.0	

in Table 8.3. The maximum temperature of the vessel and temperature of input water are not provided.

The Wave 6000/300 Tandem system (maximum working pressure of 600 MPa) from Hiperbaric is the biggest HPP system for industrial production. In this system, two vessels and their peripherals work together, sharing the same intensifier pumping group. The operation is dephased: when one vessel is in the phase of increasing pressure, the other one is in the holding-time phase. Owing to its volume and vessel diameter (two vessels, each vessel of capacity 300 L and 300 mm in diameter), it is the most productive machine in the range.

Hiperbaric also offers integration of its HPP systems into any existing production environment:

- Automation for product handling, loading, and unloading.
- Ancillary equipment for the HPP units and peripherals such as hoppers, product customized HPP containers, and basket for loading and unloading systems, post-HPP product dryers, etc.

The 525 L is the largest size single HP vessel offered by Hiperbaric (Figure 8.3). The projection is that with the throughput of 3000 kg/h, this unit can offer the least expensive cost of HPP-processed product.

MULTIVAC HPP (http://ca.multivac.com/our-products/hpp-high-pressure-preservation.html) offers industrial single and tandem systems with the vessel volumes from 55 L up to 350 L. The maximum working pressure 600 MPa and cycle time (including 3 min pressure HT) from 7.0 to 9.0 min. Maximum treated volume of product is 1900 L/h and

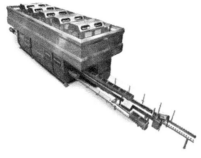

Figure 8.3 The Hiperbaric 525 L HPP unit.

Table 8.4 HPP Systems Portfolio from Multivac				
System Type	HPP-055	HPP-160	HPP-350	HPP-700 Tandem
Vessel volume, L	55	160	350	2×350
Maximum pressure, MPa	600	600	600	600
Cycle time, min (3 min HT included)	7.0–8.5	7.2–8.3	7.0–9.0	3.5–5.0
Production rate, L/h	240	1000	1900	5000
Maximum filling capacity, %	35–60	60–80	60–80	60–80
Weight, tons	17	45	55	100
Systems dimensions, m L × W	7.5×3.0	8.5×4.0	14.0×4.0	14.0×9.0

typical filling degree of 60−80%. The tandem unit HPP-700 increases process output up to 5000 L/h due to cycle time decrease to 3−5 min (Table 8.4).

MULTIVAC HPP offers processing solutions in combination with packaging using modified atmosphere (MAP) and vacuum packages. That includes package design, geometry, and format, packaging films and film laminate structure, type and formulation of product. Longfresh (AU) offers HPP toll processing using a 160 L HPP Plant from Multivac.

Bao Tou KeFa High Pressure Technology Company (http://www. btkf.com/Item/list.asp?id = 69) from China offers industrial and lab scale HPP units. The range of vessel volumes varies from 30 to 200 L at maximum pressure of 600 MPa. The units are built using horizontal layout with the installation size: 5000 (L) × 2500 (W) × 1600 (H) (mm). The water is used as pressure medium and working temperature

Figure 8.4 Bao Tou KeFa HP vessel.

can achieve 80°C. The company offers single, quos (2 × 200 L and 2 × 300 L) and quartos systems (4 × 200 L and 4 × 300 L) to increase production output. The company employs double heat-shrink sleeve and autofrettage technology that assures the maximum in pressure-vessel safety at lower weights, thus reducing facility costs (Figure 8.4).

Fresher Evolution HPP partnered with All Natural Freshness, (MI, USA) and Bao Tou KeFa is the newest HPP equipment manufacturers in North America (http://allnaturalfreshness.com/fresher-evolution-hpp-equipment-nondomestic/#sthash.d6pPbdkH.dpuf).

Fresher Evolution adapted the autofrettage vessel technology and designs key components including the material handlings, loading, and unloading baskets along with basket pallet solutions for transport to/ from tolling centers (Figure 8.5). In the moment HPP vessels with the volume of 175, 350, and 525 L are offered at the working pressure of 600 MPa. At a 70% fill rate, it is projected that the 175 L Fresher Evolution KeFa HPP machine can process approximately 27,256 lb/ 12,363 kg/1481 gallons a day or roughly 136,283 lb/61,817 kg/7407 gal-lons a week assuming a 70% machine operating time, 8 h shifts per day, and 5 days of operating a week.

In summary, four main HPP equipment manufacturers can be iden-tified around the world in the United States, Spain, Germany, and China. The vertical and horizontal orientation of HPP vessels are the

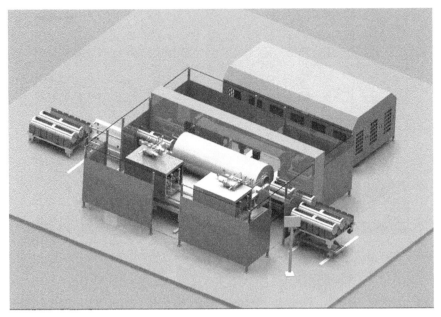

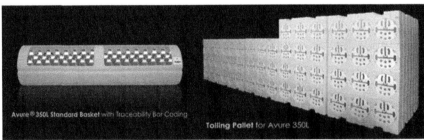

Figure 8.5 Fresher Evolution HPP solutions: HPP vessel, baskets, and tolling pallets.

main designs currently used. The volume of commercial HPP vessels varies from 55 to 525 L with a filing capacity of 60–80% that can run approximately 100,000 cycles at 600 MPa. For comparison reasons, it should be noted that the evaluation of cycle time or total cycle time differs among manufacturers that consequently may affect the output evaluation. Hiperbaric includes loading and unloading times along with CUT, CD, and HT into the total cycle time. Avure excludes HT from the total cycle, whereas Multivac includes the HT into the cycle calculation.

The HPP equipment has to comply with the most national and international directives, rulings, and standards such as CE Certificate for Pressure Equipment, 97/23/CE Directive, and The American Society of Mechanical Engineers Certificate of Authorization ASME Boiler and Pressure Vessel Code, Section VIII, Div. 3

8.5 INITIAL PURCHASING COSTS, MAINTENANCE, AND OPERATION COSTS

Commercial-scale high-pressure-processing systems cost approximately $300,000 to over $ 3.0 million, depending on the equipment capacity and the extent of automation. Currently, HPP treatment costs are quoted as ranging from 4 to 10 cents/lb. The critical factors that impact process cost and throughput are the vessel size, the pumping power, application, and processing conditions such as the cycle time, process pressure and packaging, the level of automation, and most importantly the ability of the operators and maintenance teams run an efficient operation.

Calculations of processing costs have been done taking account of amortization of the equipment cost over 5 years, assuming 3000 working days per year and 16 h per day; operating costs; wear and tear of parts; and utilities. Amortization of equipment is responsible for about 60% of the processing cost, wear and tear of parts represents 36%, energy represents less than 4%, and the cost of water consumption is negligible. These processing costs should be increased by 10−40% to include labor costs, depending on the level of automation; this includes loading products into baskets and drying after processing. An economic model of the 215 L Avure machine with two product loads is given in Table 8.5.

As the demand for HPP equipment grows, innovation is expected to reduce capital and operating costs further. In general, during the last few years, the cost of treatment per kilogram of HPP-processed products has improved. Capacities have increased owing to design improvements and optimized loading and unloading systems. The production cost of a process must be lower than the value added to the product. The value added by HPP processing can be measured in terms of higher product quality, increased product safety, and a longer product shelf-life. These issues can translate further into reduced transportation, storage, insurance and labor costs, and into consumer convenience and enhanced safety. Food companies must be able to

Table 8.5 Economic Model for the Avure 215 L HPP Machine		
	250#/Cycle	300#/Cycle
Vessel price (ea.)	$1,450,000	$1,450,000
Employees per shift	2	2
Labor cost per hour ($13.00/h burdened)	$26.00	$26.00
Energy cost/hour ($0.045/kWh × % pressure up time)	$7.83	$7.83
Pounds per cycle	250	300
Cycles per hour	8	8
Percent operating time	94%	94%
Average spare parts per cycle	$3.50	$3.50
Annual cost for HPP equipment	$150,000	$150,000
Annual depreciation cost for conveyors and automation	$35,000	$35,000
Annual depreciation cost for building	$4000	$4000
Annual spare parts cost	$157,920	$157,920
Annual labor cost	$156,000	$156,000
Annual electricity cost	$46,953	$46,953
Total annual cost	$549,873	$549,873
Cost per pound—HPP equipment, conveyors, and automation	$0.0164	$0.0137
Depreciation cost per pound—building	$0.0004	$0.0003
Spare parts cost per pound	$0.0140	$0.0117
Labor cost per pound	$0.0138	$0.0115
Electricity cost per pound	$0.0042	$0.0035
Total cost per pound	$0.0487	$0.0406

make a realistic cost–benefit analysis of the potential rewards of investment in HPP processing. The value of HPP processing in terms of increasing food safety assurance may alone be sufficient to justify an investment in some cases.

8.6 ENERGY CONSIDERATION

Most of the energy of the HPP process is consumed by the electric motor used in the pump-intensifier systems during the compression phase to operate the vessel at a desired pressure under which suitable reactions are triggered. To date, the energy required to compress the vessel is mostly lost. However, recovery of the compression energy can be achieved by synchronization of the compression and decompression phase in twin vessel systems, where up to half of the decompression energy can be used to compress the vessel that is at atmospheric pressure. Energy required for chilling or preheating the food product prior to pressure treatment is another potential source of energy utilization.

Vessel filling efficiency plays an important role in energy requirement per food product unit. Because of this, it is beneficial to maximize the product load in the vessel to fully utilize compression energy (Figure 8.6). Otherwise, the compression energy that is used to compress the pressure-transmitting fluid is largely wasted.

By comparing specific energy inputs of a thermal and a combined thermal and HPP sterilization process, Toepfl et al. (2006) estimated that the specific energy input required for the sterilization of cans can be reduced from 300 to 270 kJ/kg when a high pressure is applied. In the case of high-pressure processing, a compression energy recovery rate of 50% can be estimated when a two-vessel system or a pressure storage system is used. Making use of energy recovery, a specific energy input of 242 kJ/kg will be required for sterilization, corresponding to a reduction of 20%. Among thermophysical properties, food compressibility, thermal expansion coefficient, density, and specific heat capacity play an important role in HPP and effect thermodynamics and energy consumption.

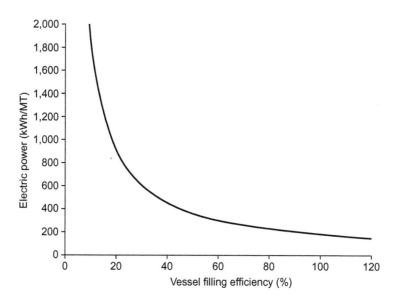

Figure 8.6 Estimated specific energy consumption in dependence of the vessel loading efficiency with food product.

8.7 HPP SYSTEM INSTALLATION CONSIDERATIONS

When planning to add HPP system to food operation, there are many considerations beyond the cost of the system that need to be planned.

Some **research** should be done to choose a manufacturer to understand all the requirements—from ordering the unit to ordering spare parts. The availability of spare parts is important consideration to avoid of undesirable delays.

System size. The area where the HPP system is to be placed needs to be large enough to house the unit and properly maintain it. The considerations should be given to the location of the intensifiers, as they are typically installed above the HPP unit. The number of intensifiers defines the cycle time and productivity. Weight considerations are also important, as the concrete floor directly under the unit must be at least 14 in thick to support the equipment's weight. The HPP vessel and the external frame is extremely heavy meaning that special equipment must be available with the size and ability to lift and push these pieces. Also, with this extreme weight, the floor needs to be protected with steel plates to help distribute the pressure and prevent the floor from crushing.

Water quality. The water that is used to fill the HPP vessel needs to be treated properly. The concern is that at such high pressures, the water may react negatively with the steel and weaken the vessel. The plant water needs to be tested to determine the hardness and if there are any chemicals present. If the hardness is too high or there are other negative factors, the proper water treatment will be required. The water test may also include chlorine, iron, hardness, connectivity, TDS (total dissolved solids). The goal for hardness is $1-25$ (mg/L). There is also a minimum (41 F) and maximum temperature (86 F) for the water.

Maintenance. The maintenance costs to operate the unit also need to be taken into account when considering an investment in HPP. Having a maintenance technician will add to the capability to service unit all the time.

Electricity cost. There is a significant amount of electricity that is used by HPP system.

Cycle time. The cycle time or throughput of the unit is another thing that must be considered in the maintenance and production planning. The HPP cycle time is affected by the required pressure and dwell-time for that particular product. The number of intensifiers in service will define CUT to achieve the desired pressure.

8.8 LAB SCALE AND PILOT PLANT HPP UNITS

In total, nine manufacturers of HPP equipment were identified that can make and offer lab scale machines. Information on laboratory and pilot HPP machines and equipment manufacturers is summarized in Table 8.6. The range of process pressure in the lab machines varies from 600 up to 981 MPa with a maximum pressure of 1400 MPa in Resato (The Netherlands) HPP multiple vessel units. The range of process temperature varies from −20°C up to 110°C allowing developing processes in combination with high or low temperatures. Only a few machines are equipped with thermal probes. However, from machines description, it is difficult to conclude whether the temperature of treated product can be measured and controlled under pressure conditions. The volume size of the vessels varies in the range of 0.03−3.5 L meaning that majority of food products cannot be HPP treated in their final commercial packages and corresponding mass and volume. Usually the food samples need to be repacked in the smaller size plastic bags.

There are a very limited number of studies that investigated and reported scalability issues in HPP process development. The lab or pilot scale machines were used in the majority of the published reports. There is a lack of equipment description and technical details. Usually only the brand name and the model are provided. The variation between HPP treatments using lab/pilot or commercial-scale units can be derived based on the following differences:

- Pressure-transmitting medium
  - Water is used in commercial machines
  - Water, water−glycol mixtures, and other fluids are used in lab scale machines
- Temperature of pressure-transmitting medium
- Rate of pressurization and depressurization
- Drop of pressure during HT
- Food package weight, size, and shape
  - smaller repackaged products are used in the pilot studies
- Sample temperature
- Cooling methods.

Table 8.6 Technical Characteristics of Laboratory Scale HPP Equipment

Company	Vessel Volume, L	Operational Parameters			Weight, kg	Dimensions, m	Comments, Power
		Pressure, MPa	Min Temp. °C	Max Temp, °C		W × L × H	
Avure Technologies, USA	2	690	0	90	1800	0.37	Water,
						1.83	3 ph
						1.47	2 thermal probes
Bao Tou KeFa High Pressure Technology Co. Ltd, China	3−5	600		80	NA	1.3	Water
						1.5	3−5 kW
						1.6	
Kobelco, Japan	0.565	686−981	20	80	NA	1.3	NA
						1.1	
Dr. Chef						1.5	
UHDE, Germany	ID 120 mm	600	0	95	NA	NA	Customized solutions
Stansted, UK	0.03	900	−20	90			
	0.30	900	−20	90			9 models
	0.675	900	−20	110			
	2.0	900	−20	110			Max
	2.8	900	−20	110			1400 MPa
	3.5	900	−20	110			
Dustec, Germany						1.1	3 thermal probes
	1.0	650	20	90	2000	2.9	
		Design pressure 700				2.1	
Resato HPP, The Netherlands	Multivessel System	Up to 1400 MPa					Offers H Systems and components
Elmhurst Research, NY, USA	Up to 100	700	40	50		1 mln cycles	Website was modified in 2007
High Pressure Dynamics, OH, USA							New company
EPSI, Belgium	Lab, pilot, and commercial HPP cold and hot presses						

CHAPTER 9

Regulatory Status Update

The regulatory status of HPP technology and the requirements for the safety evaluation of HPP process in the United States, Canada, EU, New Zealand, and Australia are discussed in the chapter.

9.1 UNITED STATES

HPP technology is accepted around the globe. The regulatory requirements within the United States for HPP-processed foods are divided between the FDA and USDA depending on the product type. Currently each product is considered on an individual basis to ensure the appropriate risks have been identified and mitigated by HPP. Yet, significantly foods processed by HPP are not considered novel and hence do not need to go through premarket approval. However, there is a need to demonstrate that a designated HPP treatment achieves a target log reduction of pathogens of concern. With fruits- and vegetable-based products, the pathogens of concern are *Salmonella*, *L. monocytogenes*, and *E. coli* O157:H7. The FDA designate that an HPP process requires to achieve a 5-log CFU reduction. In the United States, the USDA's Food Safety and Inspection Service (FSIS) has approved HPP to reduce microbial activity in RTE meat products such as marinated meats, sliced ham, turkey and chicken cold cuts, as well as cured meat products.

US FDA and USDA have approved HPP as a postpackage pasteurization technology for the manufacture of shelf-stable high acid foods and pasteurized low acid food products, and developed guidelines and regulations for those products (21CFR §114 and 21CFR §113). In 2009, the FDA approved a petition for the commercial use of a PATS process in the production of low acid foods, which was filed initially for processing in a 35 L high-pressure sterilization vessel.

9.2 CANADA

Health Canada (Novel Food Decisions, available on Health Canada's web site at http://www.novelfoods.gc.ca) issued no-objection decisions

Adapting High Hydrostatic Pressure for Food Processing Operations. DOI: http://dx.doi.org/10.1016/B978-0-12-420091-3.00009-1

about RTE meats and poultry that have been treated by HPP for the control of *L. monocytogenes* in 2006; about meat-containing entrees, meat-containing salads, and meat products in 2006; and about apple sauce and apple sauce/fruit blends in 2004.

Health Canada's Guidance for Industry on Novelty Determination of High Pressure Processing (HPP)-Treated Food Products clarifies the conditions under which an HPP-treated food is novel and therefore requires premarket assessment prior to sale in the Canadian marketplace. Any food treated with HPP for a different purpose than those approved by Health Canada, any food not previously treated with HPP, and/or any HPP-treated food that is subjected to a new set of treatment conditions (pressure, HT cycle, etc.) are considered to be novel foods as defined under Division 28 of Part B of the Food and Drug Regulations. In essence, the food is considered novel in relation to traditional counterparts as the process has caused the food to undergo a major change. A major change places the food outside the accepted limits of natural variations for that food with regard to: the composition, structure or nutritional quality of the food or its generally recognized physiological effects; the manner in which the food is metabolized in the body; or the microbiological safety, the chemical safety; or the safe use of the food.

In a premarket novel food notification, the manufacturer of a novel HPP-treated food product must submit sufficient data to Health Canada on the nutritional, chemical, toxicological, allergenic, and microbiological effects of the HPP treatment in order to demonstrate the safety of the HPP-treated food, or a sound scientific rationale for the exclusion of specific analyses.

The safety of packaging materials is controlled under Division B.23 of the Food and Drug Regulations. The Food Packaging Materials and Incidental Additives Section within the Food Directorate's Bureau of Chemical Safety is able to determine the acceptability of packaging materials that can be used for HPP-treated products. Food manufacturers of HPP-treated products can request that their packaging materials provider obtain a letter of no-objection from the Food Packaging and Incidental Additives Section regarding the acceptability of their packaging materials for HPP treatment. Further information on the regulation and premarket assessment of packaging materials can be found on Health Canada's Packaging Materials webpage. In addition, a complete list of all packaging materials accepted by the Canadian Food Inspection Agency (CFIA) including those acceptable for HPP treatment is available on CFIA's web site.

9.3 EUROPEAN UNION

The European Commission adapted existing legislation on novel foods to products processed by HPP (EC258/97; 2002). HPP is still an emerging technology in Europe and has yet to find widespread application in commercial processing. There are several underlying factors for the lack of growth of HPP within the EU that relates to consumer reluctance to pay a premium for pressure pasteurized products. In addition, food manufacturing within the EU is composed of small and medium size enterprise (SME) do not have the resources to invest in HPP or seek the premarket approval from regulatory agencies. A further factor is the status of HPP-processed foods under the Novel Foods Regulation.

The Novel Foods Regulation was implemented in 1997 to provide premarket evaluation of new "novel" products. To date, only a narrow range of products (fruit-based preparations) have been approved by the EU commission. The conditions of approval, released in 2000, were to define the processing parameters and ensure measures are in place to prevent germination of *C. botulinum*. Importantly the commission ruled that HPP was a direct alternative in the thermal pasteurization treatment of 85°C for 10 min that was traditionally used for fruit-based products with a shelf-life of greater than 10 days.

The EU is composed of member states that can act independently on food policy rather than being directed by the European Commission. In this respect, the Spanish Food Safety Agency (AESA) and the UK Food Standards Agency (FSA) permitted sale of HPP products without approval given neither agency considered high pressure a novel process, thereby did not produce novel foods. It should be noted that the removal of the "novel food" designation was limited to high acid fruit and vegetables with a shelf-life of less than 21 days. Yet, other member states had reservations of HPP given there were significant knowledge gaps with respect to microbiological issues, in addition to allergenicity and nutritional aspects. However, it was successfully argued that although HPP is a novel process, it does not produce novel foods, or changed the product composition. Therefore, provided the HPP process was sufficient to produce a product that satisfies the microbiological criteria and equipment used complies with the Pressure Equipment Directive (97/23/EC). As a consequence, there are HPP-processed products available in a number of member states (United Kingdom, Holland, and Czech Republic). A meat processor in Germany produces an HPP meat product although this is for export only to the United States.

Currently there is no definitive regulations in the EU with respect to if HPP-processed foods are considered novel by all member states. Yet, it should be noted that the HPP sector in Europe is still developing with less than half the capacity compared to North America.

9.4 NEW ZEALAND AND AUSTRALIA

Similar to Mexico, Spain, and Korea, the primary HPP-processed products available in New Zealand and Australia are avocado products. The main motivation for applying HPP in avocado production is to reduce microbial loading and inactivate spoilage enzymes (polyphenol oxidase and lipidoxigenase) thereby increasing shelf-life. In terms of pathogen reduction, the New Zealand Food Safety Authority has issued draft guidelines on the application of HPP as an intervention step. Similar to the United States, New Zealand has adopted an outcome-based approach whereby an HPP process must achieve a designated log reduction of the most relevant vegetative pathogen. The specific pathogen and log reduction requirement must be defined by the operator although a 2-log reduction was quoted in the guide. The guide also takes regulatory elements from the EU with respect to HPP process achieving the microbiological criteria for a designated product. For example, the absence of *L. monocytogenes* in RTE deli meats. A further requirement stipulated in the New Zealand guide is that HPP products should have additional hurdles to prevent the outgrowth of bacteria endospores within the expected shelf-life of the product and storage conditions. Other requirements in the New Zealand guide are in line with the EU and the United States with respect to HPP unit specifications in terms of monitoring facilities (pressure and temperature), in addition to safety standards, as designated by Pressure Equipment Directive.

The proposed guidelines for acceptance of HPP products also propose the criteria for undertaking validation trials which are more defined compared to those published by other regulatory bodies. Yet, with the large knowledge gaps that exist with regard to HPP, the recommendations for validation studies are vague and primarily left to the operator to justify. An interesting feature of the recommendations is the acceptance of previously published results and modeling to justify the HPP conditions. This is a departure from the "Novel Foods" standards where a case-by-case evaluation is required. In the case when challenge trials are performed, the log count reduction achieved is reflective of the expected levels on raw materials which is a departure from the standard 5-log reduction of relevant pathogens required in North America. A further feature of validation trials is the inclusion of shelf-life studies that must go beyond the expected time period proposed by the operator.

CHAPTER 10

Conclusions

10.1 HPP COMMERCIALIZATION AND GAPS IN TECHNOLOGY TRANSFER FOR FOOD APPLICATIONS

HPP-processed products such as deli meats, shellfish, fruits, salsa, and guacamole are commercially available in the United States, Canada, Europe, Australia, New Zealand, and Japan. For most commercial products, HPP has been applied as a final preservation step after packaging, to extend the shelf-life of the product or to enhance microbial safety through pasteurization. As example, HPP is applied to packaged meats for additional microbial safety assurance owing to the risk of postprocessing contamination from pathogens such as *L. monocytogenes*. However, products must still be stored at refrigeration temperatures. Fruit and vegetable cold-pressed juices are the fastest growing product market, whereas pressure treatment of vegetable products is limited, owing to their relatively high pH and the possible survival of pathogenic spore-forming microorganisms and texture issues. In addition to its use as a preservation step, HPP technology is being applied for raw-meat separation in seafood products, texture modification, and explored as a method for pressure-assisted freezing and thawing and a new processing solution for salt reduction and modification of the product formulation.

The HPP units are commercially available both vertical and horizontal with internal single vessel volumes ranging from 30 to 525 L. Four existing HPP equipment manufacturers have been identified in the United States, EU, and China. HPP technology is accepted around the globe including the United States, Canada, EU, New Zealand, and Australia. Owing to the complicated intrinsic and extrinsic factors, the HPP conditions need to be verified and risk analysis needs to be conducted case by case to ensure the required microbial reduction. HPP thermally assisted sterilization processes (PATS) for prepacked low acid foods are not available commercially, although they have been approved by the US FDA.

Adapting High Hydrostatic Pressure for Food Processing Operations. DOI: http://dx.doi.org/10.1016/B978-0-12-420091-3.00010-8

The biggest obstacle for HPP commercialization is the initial capital investment required. The actual cost of operating an HPP plant depends on many factors, ranging from the operating pressure, cycle time, and product geometry to the labor skills available and energy costs.

The establishment of criteria for HPP pasteurization and sterilization processes requires optimization of the process pressure, temperature, and time to inactivate target pathogenic and spoilage-causing bacteria based on knowledge of the behavior of the food under pressure. Even though a first-order destruction model can be applied to microorganisms subjected to HPP, it is observed that as the HPP process time progresses, the rate of microbial inactivation decreases and results in a survival curve with a "pressure-resistant tail." The phenomenon of a temperature increase in foods due to adiabatic compression plays a role in the establishment of the conditions for an HPP process. The magnitude of the temperature increase needs to be taken into account that is dependent on the product composition and the product initial temperature. A further requirement for validation trials is the definition of processing parameters in terms of product and processing temperatures, in addition to pressure. The cycle time must also be defined along with compression and decompression rates. The type and integrity of packaging should be provided.

Innovative research of HPP for food and biotech applications has grown worldwide aiming to overcome challenges and improve efficacy of treatment and generate new knowledge in this area. Prerecorded online courses and lectures offered by Novel Food Sciences that discuss in detail fundamentals, engineering, and packaging aspects and recent developments in the state of the art of HPP technology for foods can be downloaded at http://novel-food-sciences.com/iclasses/indexengineering.php.

REFERENCES

Ananth, V., Dickson, J.S., Olson, D.G., Murano, E.A., 1998. Shelf-life extension, safety and quality of fresh pork loin treated with high hydrostatic pressure. J. Food Prot. 61 (12), 1649–1656.

Bajovic, B., Bolumar, T., Heinz, V., 2012. Quality considerations with high pressure processing of fresh and value added meat products. Meat Sci. 92 (3), 280–289.

Barbosa-Cánovas, G.V., Juliano, P., 2008. Food sterilization by combining high pressure and heat. In: Gutierrez-López, G.F., Barbosa-Cánovas, G., Welti-Chanes, J., Paradas-Arias, E. (Eds.), Food Engineering: Integrated Approaches. Springer, New York, NY, pp. 9–46.

Bridgman, P.W., 1912. Water. In: The Liquid and Five Solid Forms, under Pressure Proceedings of the American Academy of Arts and Sciences, pp. 441–558.

Bridgman, P.W., 1931. The Physics of High Pressure. G. Bells and Sons, London.

Buckow, R., Heinz, V., 2008. High pressure processing—a database of kinetic information. Chemie Ingenieur Technik 80, 1081–1095.

Bull, M.K., Steele, R.J., Kelly, M., Olivier, S.A., Chapman, B., 2010. Packaging under pressure: effects of high pressure, high temperature processing on the barrier properties of commonly available packaging materials. Innov Food Sci. Emerg. Technol. 11 (4), 533–537.

Crehana, C.M., Troya, D.J., Buckley, D.J., 2000. Effects of salt level and high hydrostatic pressure processing on frankfurters formulated with 1.5 and 2.5% salt. Meat Sci. 55, 123–130.

Farkas, D., Hoover, D.G., 2000. High pressure processing. J. Food Sci. Suppl. 65, 47–64.

Gervilla, R., Capellas, M., Ferragut, V., Guamis, B., 1997a. Effect of high hydrostatic pressure on *Listeria innocua* 910 CECT inoculated into ewes' milk. J. Food Prot. 60 (1), 33–37.

Harvey, A.H., Peskin, A.P., Klein, S.A., 1996. NIST/ASME Standard Reference database 10, Version 2.2, National Institute of Standards and Technology, Boulder, CO.

Hendrickx, M.E.G., Knorr, D.W., 2002. Ultra High Pressure Treatments of Foods. Springer, ISBN 978-0-306-47278-7.

Herdegen, V., 1998. Hochdruckinaktivierung von Mikroorganismen in Lebensmitteln und Lebensmittelreststoffen. Munich, Germany (PhD Dissertation): Technische Universitat Munchen.

Hoover, D.G., Metrick, C., Papineau, A.M., Farkas, D.F., Knorr, D., 1989. Biological effects of high hydrostatic pressure on food microorganisms. Food Tech. 43 (3), 99–107.

Juliano, P., Koutchma, T., Sui, Q.A., Barbosa-Canovas, G.V., Sadler, G., 2010. Polymeric-based food packaging for high pressure processing. Food Eng. Rev. 2 (4), 274–297.

Knoerzer, K., Buckow, R., Versteeg, C., 2010. Adiabatic compression heating coefficients for high pressure processing—a study of some insulating polymer materials. J. Food Eng. 98 (1), 110–119.

Knorr, D., Schlueter, O., Heinz, V., 1998. Impact of high hydrostatic pressure on phase transitions of foods. Food Technol. 52, 42–45.

Koutchma, T., Song, Y., Setikaite, I., Juliano, P., Barbosa-Canovas, G.V., Dunne, C.P., et al., 2010. Packaging evaluation for high-pressure high-temperature sterilization of shelf-stable foods. J. Food Process Eng. 33 (6), 1097–1114.

Lambert, Y., Demazeau, G., Largeteau, A., Bouvier, J.M., Laborde-Croubit, S., Cabannes, M., 2000. Packaging for high-pressure treatments in the food industry. Packag. Technol. Sci. 13 (2), 63–71.

Linton, M., Patterson, M.F., 2000. High pressure processing of foods for microbiological safety and quality. Acta Microbiol. Immunol. Hung. 47 (2–3), 175–182.

Lowder, A.C., 2013. Addressing Sodium Reduction and Pathogen Internalization in Non-Intact Whole Muscle Beef: Evaluation of Dehydrated Collagen and Hydrostatic Pressure as Impact Technologies. A dissertation submitted to Oregon State University.

Ma, H., Ledward, D.A., 2013. High pressure processing of fresh meat— Is it worth it? Meat Sci. <http://dx.doi.org/10.1016/j.meatsci.2013.03.025>.

Mariappagoudar, P., 2007. Effect of Water Activity, Solute Type and Temperature on Inactivation of L. innocua during High Pressure Processing. MS thesis, IIT, Chicago.

Min, S., Sastry, S.K., Balasubramaniam, V.M., 2010. Compressibility and density of select liquid and solid foods under pressures up to 700 MPa. J. Food Eng. 96, 568–574.

Min, S.K., Sastry, S.K., 2013. In-situ pH measurement of selected liquid foods under high pressure. Innov. Food Sci. Emerg. Technol. 17 (2013), 22–26.

Min, S.K., Samaranayake, C.P., Sastry, S.K., 2011. In-situ measurement of reaction volume and calculation of pH of weak acid buffer solutions under high pressure. J. Phys. Chem. B 115, 6564–6571.

Murano, E.A., Murano, P.S., Brennan, R.E., Shenoy, K., Moreira, R.G., 1999. Application of high hydrostatic pressure to eliminate Listeria monocytogenes from fresh pork sausage. J Food Prot. 62 (5), 480–483.

Mussa, D.M., Ramaswamy, H.S., Smith, J.P., 1999. High-pressure destruction kinetics of Listeria monocytogenes on pork. J Food Prot. 62 (1), 40–45.

Palou, E., Pez-Malo, A., Barbosa-Cánovas, G.V., Welti-Chanes, J., Swanson, B.G., 1997. Effect of water activity on high hydrostatic pressure inhibition of Zigosacharomyces bailii. Lett. Appl. Microbiol. 24, 417–420.

Parish, M.E., 1998. High pressure inactivation of Saccharomyces cerevisiae, endogenous microflora and pectinmethylesterase in orange juice. J. Food Protect. 18 (1), 57–65.

Patazca, E., Koutchma, T., Balasubramaniam, V.M., 2007. Quasi-adiabatic temperature increase during high pressure processing of selected foods. J. Food Eng. 80, 199–205.

Patterson, M.F., 2005. Microbiology of pressure-treated foods. J. Appl. Microbiol. 98 (6), 1400–1409.

Patterson, M.F., Kilpatrick, D.J., 1998. The combined effect of high hydrostatic pressure and mild heat on inactivation of pathogens in milk and poultry. J. Food Prot. 61 (4), 432–436.

Patterson, M.F., Quinn, M., Simpson, R., Gilmour, A., 1995. Sensitivity of vegetative pathogens to high hydrostatic pressure treatment in phosphate buffered saline and foods. J. Food Protect. 58 (5), 524–529.

Ponce, E., Pla, R., Mor-Mur, M., Gervilla, R., Guamis, B., 1998. Inactivation of Listeria innocua inoculated in liquid whole egg by high hydrostatic pressure. J Food Prot. 61 (1), 119–122.

Ramaswamy, R., Balasubramaniam, V.M., Sastry, S.K., 2005. Properties of food materials during high-pressure processing. Encyclopedia of agricultural, food and biological engineering.

Rasanayagam, V., Balasubramaniam, V.M., Ting, E., Sizer, C.E., Anderson, C., Bush, C., 2003. Compression heating of selected fatty food substances during high pressure processing. J. Food Sci. 68 (1), 254–259.

Reddy, N.R., Solomon, H.M., Fingerhut, G., Balasubramaniam, V.M., Rhodehamel, E.J., 1999. Inactivation of Clostridium botulinum types A and B spores by high-pressure processing. IFT

Annual Meeting: Book of Abstracts. National Center for Food Safety and Technology, Illinois Institute of Technology, Chicago, IL, p. 33.

Rovere, P., Carpi, G., Dall'Aglio, G., Gola, S., Maggi, A., Miglioli, L., et al., 1996a. High-pressure heat treatments: evaluation of the sterilizing effect and of thermal damage. Ind. Conserve. 71, 473–483.

Samaranayake, C.P., Sastry, S.K., 2013. In-situ pH measurement of selected liquid foods under high pressure. Innov. Food Sci. Emerg. Technol. 17, 22–26.

Setikaite, I., Koutchma, T., Patazca, E., Parisi, B., 2009. Effects of water activity in model systems on high-pressure inactivation of *Escherichia coli*. Food Bioprocess. Technol. V2 (2), 213–221.

Smelt, J.P.P.M., Hellemons, J.C., 1998. High pressure treatment in relation to quantitative risk assessment. VTT Symposium. Technical Research Centre of Finland, Espoo, pp. 27–38.

Syed, Q.A., Buffa, M., Guamis, B., Saldo, J., 2014. Effect of compression and decompression rates of high hydrostatic pressure on inactivation of *Staphylococcus aureus* in different matrices. Food Bioprocess. Technol. 7, 1202–1207.

Ting, E., Balasubramaniam, V.M., Raghubeer, E., 2002. Determining thermal effects in high-pressure processing. Food Technol. 56, 31–35.

Toepfl, S., Mathys, A., Heinz, V., Knorr, D., 2006. Review: potential of high hydrostatic pressure and pulsed electric fields for energy efficient and environmentally friendly food processing. Food Rev. Int. 22 (4), 405–423.

Zook, C.D., Parish, M.E., Braddock, R.J., Balaban, M.O., 1999. High pressure inactivation kinetics of *Saccharomyces cerevisiae* ascospores in orange and apple juice. J. Food Sci. 64 (3), 533–535.

Printed and bound by CPI Group (UK) Ltd, Croydon, CR0 4YY

03/10/2024

01040427-0010